Locally multiplicatively-convex topological algebras

by Ernest A. Michael[1]
University of Chicago and The Institute for Advanced Study

Introduction. Since the appearance of Gelfand's fundamental "Normierte Ringe" [12][2] in 1941, the theory of Banach algebras has undergone considerable development, but very little effort has been made to extend this theory to more general topological algebras. A similar situation in the theory of Banach spaces prompted Kolmogoroff and von Neumann to introduce the notion of a locally convex space; the purpose of this paper is to generalize Banach algebras in an analogous fashion, by introducing and studying the concept (first considered by R. Arens in [2]) of a locally m-convex algebra. (Precisely in what sense these generalizations are analogous is brought out in Proposition 2.7.) Throughout this paper, we shall freely use the terminology and the results of the theory of Banach algebras ([12], [18], [24]) and the theory of locally convex spaces ([9], [22], [10]), and it is therefore desirable that the reader have a working knowledge of both these theories.[*]

A locally m-convex algebra (Definition 2.1) is a topological algebra over the complexes, with a basis for the neighborhoods of the origin consisting of sets U which are convex, symmetric, and such that $UU \subset U$. (We have restricted ourselves to complex algebras throughout this paper in order to avoid extensive ramifications, but it seems clear that the real case offers no new difficulties beyond those already encountered in the theory of real Banach algebras.) While developing our theory, it is helpful to keep in mind the following crucial properties of normed algebras:

Received by the editors February 7, 1952.

(1) This paper is essentially the author's doctoral dissertation at the University of Chicago and was written while the author was an Atomic Energy Commission predoctoral Research Fellow. The author would like to thank Professors Irving E. Segal and Irving Kaplansky for their kind help and advice during its preparation.

(2) Numbers in brackets refer to the bibliography at the end of this paper.

(*) Added in Proof: Some of our results have been obtained independently by R. Arens, and will appear in [30].

I. Every complex normed division algebra is isomorphic to the algebra of complex numbers.

II. Every complete normed algebra is an $\mathfrak{F}$-space [4], [10].

III. Every complete normed algebra is a Q-algebra (see Appendix E).

What happens to these properties if we replace "normed" by "locally m-convex"? As we shall see, (I) remains true verbatim (Proposition 2.9), and consequently provides no further difficulties and no further interest. Property (II), on the other hand, becomes false in general; we do not need it very often, but when we do, we shall simply assume it or something like it (Sections 4, 13, 14 and 15). While property (III) also generally becomes false (see Example 3.1), some of its consequences remain true, and much of our paper is devoted to deriving them; our main tool for this purpose in Theorem 5.1, which asserts that every complete, locally m-convex algebra is a projective limit of Banach algebras. In Section 13 we show that, in the presence of (II), (III) is equivalent to some apparently weaker properties.

Most of our major results are contained in Sections 5, 8, and 10-14; Sections 1-4 and 6-7 are preparatory in nature, Section 9 contains incomplete results about some non-commutative questions, and Section 15 deals with a very special problem. The appendices contain some results which we need in the main part of the paper, but whose range of application goes beyond locally m-convex algebras; Appendices A, B and C contain some original contributions, while Appendices D and E are essentially ordered collections of known results.

We take this opportunity to compile a list of references to the various unsolved problems which are mentioned in this paper. These problems are found in the following places: (1) Before Proposition 7.4. (2) Before Theorem 8.4. (3) Before Proposition 8.5. (4) Remark after Proposition 8.5. (5) Before Theorem 9.3. (6) Before Definition 11.2. (7) Before Lemma 12.3. (8) Before Lemma 13.8. (9) Remark after Theorem 14.3. Among these problems, Question 1 before Lemma 12.3 seems to be one of the most interesting and natural.

Introduction to second printing. Since the appearance of this memoir in 1952, much progress has been made towards solving the problems raised in this paper. The following summarizes what I know of such progress, and I hope that it is reasonably complete. The numbering of the problems corresponds to that at the end of the first introduction above. The references are to the supplementary bibliography at the end of this section

(1) A counterexample, which is actually a commutative F-algebra, is given by S. Rolewicz in [9].

(2), (3), (4) In [15, Theorem 4], S. Warner affirmatively solves (3) and (4), thereby also providing a partial solution to (2).

(5) This problem appears to remain open. (However, see [10] for a related result on algebras which are not locally m-convex.)

(6) An unpublished counterexample, which is actually an F-algebra and a Q-algebra, was obtained by M. Kuczma [22].

(7) (Question 1). This question, whether every multiplicative linear functional on a commutative F-algebra is continuous, is still unsolved, in spite of considerable interest and effort. Perhaps the most promising step in the direction of a positive solution was taken by R. Arens, who showed in [1] that condition I of Proposition 12.5 is always satisfied. Other published contributions have been made by A. Białynicki-Birula and W. Żelazko in [2] and by S. Warner in [12, p. 202]. Anyone working on this problem may find it rewarding to consider the following specific example: Let E be an infinite-dimensional reflexive (complex) Banach space, and B(E) the algebra of all those continuous complex functions on E which are bounded on bounded subsets of E, with the topology of uniform convergence on bounded subsets of E. Then B(E) is clearly a commutative F-algebra. If A is the closed subalgebra of B(E) generated by E*, the dual space of E, then every continuous linear functional on A is of the form $f \to f(x)$ for some $x \in E$, but I don't know whether every linear functional is of this form.

(8), (9) I know of no progress on these problems.

The following supplementary bibliography contains all the papers referred to above, as well as some others which have dealt with our subject. It is, however, not intended to be complete.

Supplementary Bibliography

1. Arens, R., Dense inverse limit rings, Michigan Math. J. 5(1958), 169-182.

2. Białynicki-Birula, A. and Żelazko, W., On the multiplicative linear functionals on the cartesian product of algebras, Bull. Acad. Polon. Sci. cl. III. 5(1957), 589-593.

3. Brooks, R. M., Boundaries for locally m-convex algebras, Duke Math. J. 34(1967), 103-116.

4. ______________. On locally m-convex *-algebras, to appear in Pacific J. Math.

5. Killam, E., The spectrum and the radical in locally m-convex algebras, Pacific J. Math. 12(1962), 581-588.

6. Mallios, A., On the spectrum of a topological tensor product of locally convex algebras, Math. Ann. 154(1964), 171-180.

7. Mitiagin, B., Rolewicz, S., and Żelazko, W., Entire functions in B_0-algebras, Studia Math. 21(1962), 291-306.

8. Neubauer, G., Zur Spektraltheorie der lokalkonvexen Algebren Math. Ann. 142(1961), 131-164.

9. Rolewicz, R., Example of semi-simple m-convex B_0-algebra, which is not a projective limit of semi-simple B-algebras, Bull. Acad. Polon. Sci. Ser. Math. Astronom. Phys. 11(1963), 459-462.

10. ______________. Some remarks on radicals in commutative B_0-algebras, Bull. Acad. Polon. Sci. Ser. Math. Astronom. Phys. 15(1967), 153-155

11. Warner, S., Polynomial completeness in locally multiplicatively-convex algebras, Duke Math. J. 23(1956), 1-12.

12. ______________. Inductive limits of normed algebras, Trans. Amer. Math. Soc. 82(1956), 190-216.

13. ______________. Weak locally multiplicatively-convex algebras, Pacific J. Math. 5(1955), 1025-1032.

14. ______________. Weakly topologized algebras, Proc. Amer. Math. Soc. 8(1957), 314-316.

15. ____________. The topology of compact convergence on continuous function spaces, Duke Math. J. 25(1958), 265-282.

16. ____________. Bornological structures, Illinois J. Math. 4(1960), 231-245.

17. Wenjen, C., On semi-normed *-algebras, Pacific J. Math. 8(1958), 177-186.

18. Żelazko, W., On the locally bounded and m-convex topological algebras, Studia Math. 19(1960), 333-356.

19. ____________. Metric generalizations of Banach algebras, Rozprawy Matematyczne 47(1965), or Yale University Notes (1964).

20. ____________. On generalized topological divisors of zero in m-convex locally convex algebras, Studia Math. 28(1966), 9-16.

21. ____________. On generalized topological divisors of zero in real m-convex algebras, to appear in Studia Math.

22. Kuczma, M., A note on a problem of E. Michael, to appear in Colloq. Math.

Conventions and notations.

1. All linear spaces and algebras (= associative linear algebras) are over the complex numbers. A topological linear space is a linear space, equipped with a Hausdorff topology which makes addition and scalar multiplication jointly continuous. A topological algebra is an algebra and a topological linear space such that ring multiplication is separately continuous. A locally convex algebra is a topological algebra and a locally convex linear space.

2. If E is a topological linear space, then E' denotes the space of continuous linear functionals on E. If $U \subset E$, then $U^{o} = \{f \varepsilon E' \mid \sup_{x \varepsilon U} |f(x)| \leq 1\}$, and if $V \subset E'$, then $V^{o} = \{x \varepsilon E \mid \sup_{f \varepsilon V} |f(x)| \leq 1\}$ (see [10]). If A is a topological algebra, then $\mathcal{M}(A) = \{f \varepsilon A' \mid f(xy) = f(x)f(y)$ for all $x, y \varepsilon A\}$, equipped with the relative topology induced on it by the topology $\sigma(A', A)$ on A'. ($\sigma(A', A)$ [10] denotes the weak topology induced on A' by A.) Furthermore, $\mathcal{M}^{-}(A)$ denotes the subspace (with the relative topology) of $\mathcal{M}(A)$ which we get by removing the zero functional from $\mathcal{M}(A)$.

3. By the cartesian product of a collection of topological algebras, we mean the full direct sum of this collection with the cartesian product topology. A subalgebra of a topological algebra is assumed to carry the relative topology.

4. In all function algebras, multiplication and addition are assumed to be pointwise. A function algebra A on a set T is separating if, whenever t_1 and t_2 are distinct elements of T, there exists an x in A such that $x(t_1) \neq x(t_2)$. If T is a topological space, then C(T) denotes the algebra of all continuous, complex-valued functions on T. If furthermore $t_o \varepsilon T$, then $C_{t_o}(T)$ denotes $\{x \varepsilon C(T) \mid x(t_o) = 0\}$. (In particular, if A is a topological algebra, then $C_o(\mathcal{M}(A))$ denotes the algebra of all continuous, complex-valued functions on $\mathcal{M}(A)$ which vanish at the zero functional 0.)

5. The terms "unit", "identity", and "identity element" mean the same thing. If A is an algebra, and I a right (resp.left) ideal in A, then a unit modulo I means an element $e \varepsilon A$ such that $(e - 1)A \subset I$ (resp. $A(e - 1) \subset I$): If $x \varepsilon A$, then "x is invertible in A" means "A has a unit, and x has an inverse in A".

6. We follow Kaplansky [18] in the use of the following terminology: quasi-inverse, quasi-regular, radical, spectrum, primitive, x o y

(= x + y + xy), and x' (= the quasi-inverse of x if x is quasi-regular). We use r.q.r. as abbreviation for "right quasi-regular", and similarly for l.q.r. and q.r. We use $\mathcal{R}(A)$ to denote the radical of an algebra A. If x is an element of the algebra A, the $\Sigma_A(x)$ denotes the spectrum of x in A (= $\{\lambda \neq 0 | -\lambda^{-1}x$ is not q.r. in A$\}$, with $\{0\}$ added unless x is invertible in A), and $\sigma_A(x)$ denotes the spectral radius of x in A (= sup $\{|\lambda| \mid \lambda \in \Sigma_A(x)\}$).

7. If A is an algebra and x ε A, then $R_x: A \longrightarrow A$ is defined by "$R_x(y) = yx$", and R_A denotes $\{R_x | x \in A\}$. Similarly for L_x and L_A.

8. If A is a (topological) algebra, then A^+ will denote the algebra which we get by adding a unit to A. A^+ is merely the vector space direct sum of A and the complex numbers (in the cartesian product topology), with multiplication defined by "$(x, \lambda)(y, \mu) = (xy + \lambda y + \mu x, \lambda\mu)$".

9. "<u>Homomorphism</u>" means "algebraic homomorphism". "<u>Isomorphism</u>" and "<u>isomorphic</u>" imply that both algebraic and topological structures are preserved; if only algebraic structure is preserved, the terms "<u>algebraic isomorphism</u>" and "<u>algebraically isomorphic</u>" will be used.

10. Let T be a topological space. Then T is a <u>Lindelöf</u> space if every covering by open sets has a countable subcovering, T is <u>σ-compact</u> if it is the union of a countable collection of compact subsets, and T is <u>countably compact</u> if every infinite subset has a limit point.

11. In the algebra A of continuous, complex-valued functions on a topological space T, we use the terms <u>compact-open</u> topology (= k-topology in [1]) and <u>topology of uniform convergence on compact sets</u> [5, pg.5,III] synonymously. (Their equivalence was established in [1, Cor.9.1] and in [5, pg.18, prop.7]). A basis for the neighborhoods of 0 in this topology is given by all sets of the form $\{x \in A | \ |x(t)| < \varepsilon$ whenever $t \in C\}$, where C is a compact subset of T, and ε > 0.

12. We use <u>neighborhood</u> of a point in the sense of Bourbaki; i.e., any set which contains an open set containing the point.

13. The letters i and j will denote indices from a not necessarily countable index set; the index set itself will usually not be mentioned explicitly.

1. Idempotent and m-convex sets.

Definition 1.1. A subset U of an algebra is called idempotent if $UU \subset U$; it is called m-convex (or multiplicatively-convex) if it is convex and idempotent.

The significance of idempotence, and its relation to convexity, can be illustrated by the corresponding properties of the "pseudonorm". To define a pseudonorm, we suppose that U is a subset of the algebra A which intersects each real, one-dimensional subspace in a proper, closed interval about 0. We then define the real valued function p on A by "$p(x) = \text{g.l.b.}\{\lambda \geq 0 \mid x \in \lambda U\}$", and it is easy to see that $U = \{x \in A \mid p(x) \leq 1\}$. The following lemma now follows from the definitions:

Lemma 1.2. Let A, U and p be as in the last paragraph. Then

a) U is convex if and only if $p(x + y) \leq p(x) + p(y)$ for $x, y \in A$.

b) U is idempotent if and only if $p(xy) \leq p(x)p(y)$ for $x, y \in A$.

c) U is symmetric (3) if and only if $p(\lambda x) = |\lambda| p(x)$ for $x \in A$ and λ complex.

Remark. If p has the properties of (a) and (c), it is usually called a pseudonorm.

Remark. The norm of a normed space (resp. algebra) has the properties of p in (a) and (c) (resp. (a), (b), and (c)), as well as the property that only the norm of 0 is 0.

As examples of m-convex sets, we now mention the whole algebra, the set consisting only of 0, and the open or closed unit sphere of a normed algebra.

Lemma 1.3. The convex hull of an idempotent set is m-convex.

Proof. Let U be idempotent, with convex hull V. Since V is certainly convex, we need only show that it is idempotent. Now if x and y are in V, then $x = \sum_{n=1}^{p} a_n x_n$ and $y = \sum_{m=1}^{q} b_m y_m$, where all the x_n and y_m are in U,

---

(3) A subset U of a linear space is symmetric if $\lambda U \subset U$ for all scalars λ such that $|\lambda| = 1$.

and the a_n and b_m are scalars between 0 and 1 such that $\sum_{n=1}^{p} a_n = \sum_{m=1}^{q} b_m = 1$. Then $xy = \sum_{n=1}^{p} \sum_{m=1}^{q} a_n b_m x_n y_m$. But $x_n y_m \in U$ for all n and m, $0 \leq a_n b_m \leq 1$ for all n and m, and $\sum_{n=1}^{p} \sum_{m=1}^{q} a_n b_m = 1$; hence $xy \in V$, which is what we had to show.

Q.E.D.

In the following lemma we summarize some other useful facts about idempotent and m-convex sets.

Lemma 1.4. *Let A be an algebra. If the subset U of A is idempotent (resp. m-convex), then so are*

a) *The image or inverse image of U under a homomorphism.*

b) *The closure of U, if A is a topological algebra.*

c) λU, *if* λ *is a scalar with* $|\lambda| \leq 1$, *and if U is symmetric.*(3)

Proof. Assertions (a) and (c) are obvious. To prove (b), let $\overline{U}$ be the closure of U in A. Then, for all $x \in U$, we have $xU \subset U$, and hence, by continuity, $x\overline{U} \subset \overline{U}$. Hence $Uy \subset \overline{U}$, so, by continuity, $\overline{U}y \subset \overline{U}$ for all $y \in \overline{U}$. In other words, $\overline{U}\,\overline{U} \subset \overline{U}$.

Q.E.D.

2. *Definition and fundamental properties of locally m-convex algebras*

In the following definition, we will simultaneously define locally m-convex topological algebras and locally convex topological linear spaces, in order to emphasize the parallelism.

Definition 2.1. A topological algebra (resp. linear space) is called *locally* m-*convex* (resp. *convex*) if there exists a basis for the neighborhoods of the origin consisting of sets which are m-convex (resp. convex) and symmetric.

For convenience, we will henceforth omit the word "topological" from "locally m-convex (resp. convex) topological algebra (resp. space)." We will also adopt the following

Convention 2.2. If A is a locally m-convex algebra (resp. locally convex space) then "*m-base* (resp. *base*) *for* A" will stand for "collection of m-convex (resp. convex) and symmetric subsets of A whose scalar multiples form a base for the neighborhoods of the origin." Similarly for "m-sub-base (resp. sub-base)."

It is clear that every locally m-convex algebra is a locally convex algebra with jointly continuous ring multiplication. The converse is false, even if the algebra is commutative, metrizable, and complete [2, Th. 2]; we do however, have (similar to [12, Satz 1])

Proposition 2.3. *Let A be a metrizable and complete locally convex algebra with sub-base $\{U_i\}$, and suppose that, for every $x \in A$ and every i, $xU_i \subset \lambda_{x,i} U_i$ for some scalar $\lambda_{x,i}$. Then A is locally m-convex.*

We will not prove this proposition now, because we will state and prove a stronger proposition (Proposition 4.3) in Section 4.

For topological algebras which are normed spaces, the property of being locally m-convex is easily interpreted. In fact, it follows from the definitions that such an algebra is locally m-convex if and only if it can be renormed as a normed algebra (in the usual sense).

We will now give some examples of locally m-convex algebras, and methods for constructing new ones out of old.

Proposition 2.4.

a) *Every normed algebra is locally m-convex.*
b) *A subalgebra of a locally m-convex algebra is locally m-convex in the relative topology.*
c) *The cartesian product of locally m-convex algebras is locally m-convex.*
d) *An algebra A, with the weakest topology making continuous each element of some separating set of homomorphisms from A to various locally m-convex algebras, is locally m-convex.*
e) *If A is locally m-convex, and if I is a closed, two-sided ideal in A, the A/I is locally m-convex.*
f) *If A is locally m-convex, then A^+ is also locally m-convex.*
g) *If A is an algebra of continuous functions from a topological space T to a locally m-convex algebra, in the topology of uniform convergence on some collection of compact subsets of T whose union is T, then A is locally m-convex.*

h) *Let p be an "integer" such that* $0 \leq p \leq +\infty$, *and let A be an algebra of complex valued function on an interval T (finite or infinite) of the real line R, such that each* $x \in A$ *has everywhere at least p continuous derivatives. Then A is locally m-convex under the topology generated by the sub-base* $\{x \in A \mid |x^{(n)}(t)| \leq 1 \text{ for } t \in C\}$, *where n ranges over all integers* $\leq p$ *if p is finite and over all integers if* $p = +\infty$, *and where C ranges over the compact subsets of T.*

i) *Let T be a set, S a collection of real valued functions on T such that* $0 \leq f(t) \leq 1$ *for* $t \in T$ *and* $f \in S$, *and B a normed algebra. Let A be an algebra of functions from T to B such that, for all* $x \in A$ *and* $f \in S$, *there exists a scalar* $\lambda \geq 0$ *such that* $\| x(t) \| \leq \lambda f(t)$ *for all* $t \in T$. *Then A becomes a locally m-convex algebra if we take as an m-sub-base the sets* $\{x \in A \mid \| x(t) \| \leq f(t) \text{ for all } t \in T\}$, *as f ranges over S.*

j) *Let E be a locally convex space, with base* $\{U_i\}$. *Then the algebra* $\mathcal{L}\{U_i, U_i\}$, *with the topology* $\tau\{U_i, U_i\}$ *(Def. A.1 and Def. A.3) is locally m-convex.*

Proof. All cases, except (f) and (h), require only a straightforward verification of Definition 2.1. Let us prove (f) and (h):

(f). Let $\{U_i\}$ be an m-base for A; then $\{V_i\}$ forms a base for A^+, where $V_i = \{(x, \lambda) \mid x \in U_i, \ |\lambda| \leq 1\}$. Now if x^+ and y^+ are in V_i, then $x^+ = (x, \lambda)$ and $y^+ = (y, \mu)$, where x and y are in U_i, and $|\lambda| \leq 1$ and $|\mu| \leq 1$. Hence $x^+y^+ = (xy + \mu x + \lambda y, \mu\lambda) \in 3V_i$, and therefore $V_iV_i \subset 3V_i$. Setting $U_i^+ = \frac{1}{3} V_i$, we see that $\{U_i^+\}$ is an m-base for A^+.

(h). Let $U_{n,C} = \{x \in A \mid |x^{(k)}(t)| \leq 2^{-n} \text{ for } k \leq n \text{ and } t \in C\}$ where n ranges over all integers $\leq p$ if p is finite and over all integers if $p = +\infty$, and where C ranges over the compact subsets of T. Then certainly $\{U_{n,C}\}$ is a base for A, and we need only show that each $U_{n,C}$ is idempotent. Now if x and y are in A, then $(xy)^{(k)} = \sum_{i=1}^{2^k} x_i y_i$, where the x_i are derivatives of x of some order $\leq k$, and similarly for the y_i. Hence if x and y are in $U_{n,C}$, then $|xy^{(k)}(t)| \leq 2^n 2^{-n} 2^{-n} = 2^{-n}$ if $k \leq n$ and $t \in C$, and hence $xy \in U_{n,C}$.

Q.E.D.

Remark. Ad (h): This can easily be generalized to differentiable functions on an open subset of Euclidean n-space.

Ad (j): We will see in Proposition 4.4 that all complete and metrizable locally m-convex algebras are isomorphic to a transformation algebra of this kind.

Section 3 will be devoted to giving specific examples of complete, locally m-convex algebras of complex-valued functions.

We now proceed to show that every locally m-convex algebra is generated, in a certain sense, by normed algebras. This fact will be the corner stone of our whole theory, for it will enable us to reduce most of our problems to similar problems about normed algebras. It also lays the foundation for Proposition 2.7 below, which "justifies" the concept of locally m-convex algebras.

Lemma 2.5. *Let A be an algebra (resp. linear space), and let U be an m-convex (resp. convex) and symmetric set which spans A. Let p be the pseudonorm associated with U, and N the null set of p. Then N is an ideal (resp. subspace) in A, and $p(x) = p(y)$ whenever $x - y \in N$. If we now define $\tilde{p}$ on A/N by $\tilde{p}(x + N) = p(x)$, then $\tilde{p}$ is a norm on A/N, which makes A/N into a normed algebra (resp. space).*

Notation 2.6. Let A be a locally m-convex algebra (resp. locally convex space) and let $\{U_i\}$ be an m-base (resp. base) for A. Then, for each i,

a) "p_i" denotes the pseudonorm on A generated by U_i.

b) "N_i" denotes the null space of p_i.

c) "$\tilde{p}_i$" denotes the norm induced by p_i on the abstract algebra (resp. linear space) A/N_i as in Lemma 2.5.

d) "A_i" denotes the normed algebra (resp. space) which we get by equipping A/N_i with the norm $\tilde{p}_i$.

e) "$\bar{A}_i$" denotes the completion of A_i.

f) "π_i" denotes the natural homomorphism from A onto A_i.

g) "x_i" denotes $\pi_i(x)$ $(x \in A)$.

Remark. Observe that π_i is continuous.

Remark. In general, the N_i are different from 0 (*e.g.*, Ex.3.1, Ex.3.2), but it may happen that they are all 0, and that all the p_i are therefore not only pseudonorms but actually norms (Ex. 3.4 and Ex. 3.5). Again, the A_i need not be complete even if A is complete (Ex. 3.4), but frequently they all <u>are</u> complete (Ex. 3.1).

We are now ready for the proposition which we promised earlier, which gives a precise meaning to the first sentence of the introduction.

Proposition 2.7. *A topological algebra (resp. linear space) is locally m-convex (resp. convex) if and only if it is isomorphic to a sub-algebra (resp. sub-space) of a cartesian product of normed algebras (resp. spaces).*

Proof. A subalgebra (resp. subspace) of a cartesian product of normed algebras (resp. spaces) is certainly locally m-convex (convex), by Proposition 2.4, (a) - (c).

Suppose, conversely, that A is a locally m-convex algebra (resp. locally convex space), and let $\{U_i\}$ be an m-base (resp. base) for A. Then, using Notation 2.6, it follows from Definition 2.1 that the topology on A is the weakest topology for which all the π_i are continuous. Consequently, A is naturally isomorphic to a subalgebra (resp. subspace) of the cartesian product of the A_i. Q.E.D.

We conclude this section with the following propositions, which are our first results about the <u>ring</u> properties of a locally m-convex algebra.

Proposition 2.8. *If A is a locally m-convex algebra, then quasi-inversion is continuous on the set of q.r. elements.*

Proof. Let $\{U_i\}$ be an m-base for A. It is certainly sufficient to show that, for all i, $p_i(x' - y')$ is small whenever $p_i(x - y)$ is small. This goes through verbatim as in the case of normed algebras, with the norm replaced by p_i.

From Proposition 2.8 we conclude the following proposition, which is true for arbitrary locally convex algebras with continuous quasi-inversion.

Proposition 2.9. *Let A be a locally m-convex algebra. Then*

a) *If $x \in A$, then $\Sigma_A(x)$ is not empty.*

b) *If A is a division algebra, then A is isomorphic to the algebra of complex numbers.*

c) *If A is commutative, and if M is a closed, regular, maximal ideal in A, then A/M is isomorphic to the algebra of complex numbers.*

Proof.

a) If A has no unit, then $0 \in \Sigma_A(x)$ by definition. If A has a unit, then (a) is proved in [17, pg.811] for arbitrary locally convex algebras with continuous inversion.

b) This follows immediately from (a).

c) This follows from (b), since every simple, commutative ring with unit is a field.

Q. E. D.

Corollary 2.10. If A is a commutative, locally m-convex algebra, then there is a 1-1 correspondence between $\mathcal{M}^-(A)$ and the collection of closed, regular, maximal ideals in A, such that every $f \in \mathcal{M}^-(A)$ corresponds to its kernel.

Proof. This follows from Proposition 2.9c, and the fact that the complexes admit no algebra automorphism other than the identity.

3. Examples of complete locally m-convex algebras of complex-valued functions. Our reason for giving only examples of function algebras is that these are the ones which we can most easily examine, and which we know most about. We will, moreover, see in Proposition 8.1 that every semisimple, commutative, and complete locally m-convex algebra is isomorphic to a function algebra; we are therefore considering a very general class of commutative algebras. For the non-commutative case, we invite the reader to consider operator algebras as described in Proposition 2.4, or to start with a non-cummutative Banach algebra, and go to work on it with the tools of Proposition 2.4.

Our reason for requiring completeness is that, just as in the case of normed algebras, this property turns out to be very important; see for instance Section 5.

All the algebras in the examples below (except Example 3.6) are special cases of the algebras of Proposition 2.4 (g) or (h), and are therefore automatically locally m-convex. Our only job will therefore be to prove that these algebras are complete. We will make frequent use of Appendix D.

Example 3.1. A is the algebra of continuous, complex-valued functions on the real line, in the compact-open tcpology.

Proof. This follows from Example D.3 and Lemma D.5.

Example 3.2. A is the algebra of entire functions on the complex plane T, in the compact-open topology.

Proof. By Example D.3 and Lemma D.5, the algebra C(T) of all continuous functions on T is complete in this topology, so we need only show that the A is closed in C(T). Let $x \varepsilon C(T)$ be in the closure of A. Pick $t \varepsilon T$, and let K be an open circle with center K. Since $\overline{K}$ is compact, there exists a sequence of entire functions which approaches x uniformly on K. Hence x is analytic in K, and hence at t.

As a generalization of Example 3.2, we have

Example 3.3. Let T be a subset of the space of n complex variables, and let A be the algebra of continuous, complex-valued functions on T which are analytic in the interior of T, in the compact-open topology.

Proof. This goes just like Example 3.2.

Example 3.4. Let A be the algebra of all complex-valued functions on the real line which vanish outside the closed interval I and which have derivatives of all orders, with the topology generated by the sub-base $\{x \varepsilon A \mid |x^{(n)}(t)| \leq 1 \text{ for } t \varepsilon I\}_{n=1}^{\infty}$.

Proof. Let $\{x_i\}$ be a Cauchy sequence in A. Then for all n, the sequence $\{x_i^{(n)}\}_{i=1}^{\infty}$ is uniformly Cauchy, and hence approaches a limit function y_n which clearly vanishes outside I. By [13, Th.8, p.103], $y_{n+1} = y_n^{(1)}$ for all n, and therefore $y_n = y_1^{(n)}$ for all n. Hence $y_1 \varepsilon A$, and $x_i \longrightarrow y_1$ in A.

Example 3.5. A is the algebra of all continuous, complex-valued functions on the real line T which vanish outside compact sets, in the topology generated by the m-base $\{x \varepsilon A \mid |x(t)| \leq m(t)\}$, where m ranges over all continuous, positive functions on T.

Remark. This is the $\mathcal{LF}$-space generated by the $\mathcal{F}$-spaces A_n, where A_n is the space of continuous, complex-valued functions on T which vanish outside the interval $[n,n]$, in the topology of uniform convergence.

Proof. The completeness follows from the second assertion in the above remark and [20, p.629]. It can also be proved directly, by proceeding as in the proof of [25, Th. I, p.68].

Example 3.6. A is the $\mathcal{LF}$-space of complex-valued functions on the real line which vanish outside a compact set, and which are infinitely differentiable. (A is the space $\mathfrak{D}$ of [25].)

Proof. That this is indeed a locally m-convex algebra follows from Example 3.5 and Theorem 15.4, but is not hard to prove directly. Completeness follows from [20, p.629] or [25, Th. I, p.68].

Example 3.7. A is the algebra of continuous, complex-valued functions on the space T of ordinals smaller than Ω in the order topology, with the compact-open topology.

Proof. This follows from Example D.3 and Lemma D.5.

Example 3.8. A is the algebra of continuous, complex-valued functions on a bounded, closed interval I of the real line, in the topology of uniform convergence on the compact and countable subsets of I.

Proof. This follows from Example D.2 and Lemma D.5.

The question of metrizability and similar properties of these examples is treated at the end of Section 4.

4. $\mathcal{F}$-algebras and their relatives.

Definition 4.1. An $\mathcal{F}$-algebra (resp. $\mathcal{F}$-space) is a metrizable and complete locally m-convex algebra (resp. locally convex space).

Remark. The above definition of $\mathcal{F}$-space is due to [10, p.65]. Banach [4, p.35] and the Polish school do not imply local convexity by the term.

It is a well known ([4] and [10]) that $\mathcal{F}$-spaces have many important properties, some of which we need in the assumptions of some of the theorems of this paper (e.g., 4.3, 4.4, 8.4, 8.7, Sec. 12, 13.6, 13.7, Sec. 14, 15.4). For the sake of generality, we will assume these properties directly, rather than make the stronger assumption that the algebra (resp. space) in question is an $\mathcal{F}$-algebra (resp. $\mathcal{F}$-space). In the following proposition, which is primarily a proposition about locally convex spaces, we will establish the relations between these various properties.

Proposition 4.2. *Consider the following properties of a locally convex algebra* A:

a) A *is an* $\mathcal{F}$-*space*.

b) A *is of the second category*.

c) A *is an* $\mathcal{LF}$-*space*.

d) A *is sequentially complete, and every linear transformation to another locally convex space which sends bounded sets into bounded sets is continuous*.

e) *Every subset of A which is convex, symmetric, and closed, and whose scalar multiples cover* A, *is a neighborhood of* 0.

f) A *is sequentially complete and a locally* m-*convex algebra, and every homomorphism to another locally* m-*convex algebra which sends bounded sets into bounded sets is continuous*.

g) *Every subset of* A *which is* m-*convex, symmetric, and closed, and whose scalar multiples cover* A, *is a neighborhood of* 0.

h) *Every compact subset of* $\mathcal{M}(A)$ *is equicontinuous*.

Among these properties, the following implications hold:

$$\begin{array}{ccccc} & & a & \longrightarrow & b & & \\ & & \downarrow & & \downarrow & & \\ c & \longrightarrow & d & \longrightarrow & e & & \\ & & \downarrow & & \downarrow & & \\ & & f & \longrightarrow & g & \longrightarrow & h \end{array}$$

Proof.

a $\longrightarrow$ b: Every complete metric space is of the second category.

a $\longrightarrow$ d and c $\longrightarrow$ d: See [10; Corollary 1, p.70, and Prop.6].

b $\longrightarrow$ e: Let U be the set discussed in (e). Then $\bigcup_{n=1}^{\infty} nU = A$, so nU has an interior for some n, and hence for all n. Since U is convex, it is therefore a neighborhood of 0.

d $\longrightarrow$ e: Assume (d), and let U be the subset of A discussed in (e). Since the scalar multiples of U cover A, the polar U° of U is weakly bounded in A'. Since A is sequentially complete, it follows from [21, Th. VI-4] and [22, Th.13], (it is possible to give a simpler proof), that every weakly bounded subset of A' is strongly bounded. Hence U° is strongly bounded. Consequently, given any bounded subset B of A, there exists a scalar $\lambda \neq 0$ such that $\lambda B \subset U$ (since $U = (U^o)^o$ by [10, Th.1]). But the second part of (d) implies, as is easily seen, that every convex and symmetric subset U of A with this property is a neighborhood of the origin.

$d \to f$ and $e \to g$: These are obvious.

$f \to g$: This goes just like $d \to e$.

$g \to h$: A subset C of A' is equicontinuous if and only if C^o is a neighborhood of the origin in A. Now if C is a compact subset of $\mathcal{M}(A)$, then C^o satisfies all the assumptions of (g), and is therefore a neighborhood of the origin.

Q.E.D.

Of the properties of Proposition 4.2, (e) and (g) are among the most useful. They apply to a wider class of spaces than the closed graph theorem [4, Th.5, pg.41], and can frequently be used instead of the latter theorem. The full force of the closed graph theorem is, in fact, only rarely necessary; in this paper we need it only in the proof of Theorem 12.7 and in Section 14. In our next two propositions, we establish some consequences of property (g) of Proposition 4.2. Proposition 4.3 generalizes Proposition 2.3, and should be compared with [12, Satz 1].

Proposition 4.3. Let A be a locally convex algebra which satisfies property (g) of Proposition 4.2. Suppose that, for some base $\{U_i\}$ for A, every $x \in A$ has the property that $xU_i \subset \lambda_{x,i} U_i$ ($\lambda_{x,i}$ a scalar depending on x and i) for every i. Then A is locally m-convex.

Proof. Since $x\overline{U}_i \subset \lambda_{x,i}\overline{U}_i$ for all i and all $x \in A$, and since $\{\overline{U}_i\}$ is also a base for A, we may assume that every U_i is closed. Let A^+ be the algebra obtained by adding a unit to A, and let $U_i^+ = \{(x, \lambda) \mid x \in U_i, |\lambda| \leq 1\}$. Then $\{U_i^+\}$ is a base for A^+, and every U_i^+ is closed. Let $x \in A$ and pick an i; then $x \in \mu_{x,i} U_i$ for some scalar $\mu_{x,i}$, and therefore $xU_i^+ \in (\lambda_{x,i} + \mu_{x,i})U_i^+$. Consequently, if $V_i = \{x \in A \mid xU_i^+ \subset U_i^+\}$, then the scalar multiples of V_i cover A. Also every V_i is m-convex, symmetric, and closed. Hence, by property (g) of Proposition 4.2, every V_i is a neighborhood of 0 in A. But $V_i \subset U_i$ (since $(0, 1) \in U_i^+$), and therefore $\{V_i\}$ is an m-base for A.

Q.E.D.

Our next proposition deals with algebras of operators, and uses concepts and terminology which are developed in Appendix A.

Proposition 4.4. Let A be a locally m-convex algebra which satisfies property (g) of Proposition 4.2. Then there exists a locally convex space E with base $\{U_i\}$ such that A is isomorphic to a subalgebra (with the relative topology) of $\mathcal{L}\{U_i, U_i\}$ with the topology $\tau\{U_i, U_i\}$. If A has a unit, we may take E = A; otherwise, we may take $E = A^+$.

Proof. Let us first suppose that A has a unit e. Let $\{U_i\}$ be an m-base for A consisting of closed sets, and let $\emptyset$ be the canonical algebraic isomorphism from A onto $L_A \subset \mathcal{L}\{U_i, U_i\}$. Taking L_A in the topology induced on it by $\tau\{U_i, U_i\}$, we must show that $\emptyset$ is a homeomorphism. If $W_i = \{L_x \varepsilon L_A \mid L_x(U_i) \subset U_i\}$, then $\{W_i\}$ is, by definition, an m-sub-base for L_A. For each i, there is a scalar $\lambda_i \neq 0$ such that $\lambda_i e \varepsilon U_i$. Hence $\emptyset(U_i) \supset (1/\lambda_i)W_i$, and therefore $\emptyset$ is open. To show that $\emptyset$ is continuous, we must show that $\emptyset^{-1}(W_i)$ is a neighborhood of 0 in A for all i. But this follows from property (g), since each $\emptyset^{-1}(W_i)$ is m-convex, symmetric, and closed in A.

If A does not have a unit, then we map A into L_{A^+}. The proof now proceeds just as above (see the proof of Proposition 4.5).

Q.E.D.

Let us conclude this section by examining the examples of Section 3 for the various properties of Proposition 4.2. Examples 3.1, 3.2, and 3.6 clearly satisfy (a). Examples 3.5 and 3.6 satisfy (c), but not (a) [26, Cor. 2, p.70]. Example 3.7 does not satisfy (g), as is seen by considering $\{x \varepsilon A \mid |x(t)| \leq 1 \text{ for } t \varepsilon T\}$; it does satisfy (h), since T is homeomorphic to $\mathcal{M}^-(A)$ (Example 7.6). Example 3.8, finally, does not even satisfy (h); in fact, I is clearly homeomorphic to $\mathcal{M}^-(A)$, and I is compact, but $\mathcal{M}^-(A)$ is not equicontinuous.

5. Complete locally m-convex algebras. Let A be a locally m-convex algebra, with m-base $\{U_i\}_{i \varepsilon I}$. We may order the index set I by putting $j \geq i$ if and only if $U_j \subset U_i$; it is clear that this ordering turns I into a directed system. If $j \geq i$, and if $x_j = y_j$, then clearly $x_i = y_i$; we may

therefore define $\pi_{ij}: A_j \longrightarrow A_i$ by " $\pi_{ij}(x_j) = x_i$", and π_{ij} is clearly a continuous homomorphism onto A_i. Consequently, π_{ij} can be uniquely extended to a continuous homomorphism $\bar{\pi}_{ij}$ from $\bar{A}_j$ into $\bar{A}_i$. Consider now the cartesian product $\prod_i \bar{A}_i$, and let $\bar{\pi}_i$ be the projection from $\prod_i \bar{A}_i$ onto $\bar{A}_i$. Following [28, p.23], we now consider $\mathcal{LP}(A) = \{x \in \prod_i \bar{A}_i \mid \bar{\pi}_{ij}(\bar{\pi}_j(x)) = \bar{\pi}_i(x)$ whenever $j \geq i\}$.

It is clear that $\mathcal{LP}(A)$(4) is a subalgebra of $\prod_i \bar{A}_i$, and that the sets $\{\mathcal{LP}(A) \cap \bar{\pi}_i^{-1}(U_i)\}$ form a base (not just a sub-base!) for $\mathcal{LP}(A)$. Consider now the canonical map $\phi: A \longrightarrow \prod_i \bar{A}_i$, defined by "$\phi(x) = \{x_i\}$". It is easy to see, and well known (and was used implicitly in the proof of Proposition 2.7) that ϕ is an isomorphism into. Observe now that, in fact, $\phi(A) \subset \mathcal{LP}(A)$; A is thus canonically isomorphic with a subalgebra of $\mathcal{LP}(A)$, and henceforth we will freely identify A with its image in $\mathcal{LP}(A)$. Using this convention, observe that if $x \in A$, then $\pi_i(x) = \bar{\pi}_i(x)$ for all i.

We now have the following fundamental theorem.

Theorem 5.1. <u>Let</u> A <u>be a complete, locally</u> m-<u>convex algebra with</u> m-<u>base</u> $\{U_i\}$, <u>and let</u> $\{\bar{\pi}_{ij}\}$ <u>and</u> $\mathcal{LP}(A)$ <u>be as above. Then</u>

a) $A = \mathcal{LP}(A)$

b) <u>If</u> $x(i) \in \bar{A}_i$ <u>for each</u> i, <u>and if</u> $\bar{\pi}_{ij}(x(j)) = x(i)$ <u>whenever</u> $j \geq i$, <u>then there exists an</u> $x \in A$ <u>such that</u> $x_i = x(i)$ <u>for all</u> i.

Proof. By our definitions, (b) follows immediately from (a). To prove (a), we need only show that A is dense in $\mathcal{LP}(A)$ (this holds even for incomplete A). Since the sets $\{\mathcal{LP}(A) \cap \bar{\pi}_i^{-1}(U_i)\}$ form a base for $\mathcal{LP}(A)$, it is sufficient to show that $\bar{\pi}_i(A)$ is dense in $\bar{\pi}_i(\mathcal{LP}(A))$ for all i. But $\bar{\pi}_i(A) = \pi_i(A) = A_i$, and $\bar{\pi}_i(\mathcal{LP}(A)) \subset \bar{A}_i$; since A_i is dense in $\bar{A}_i$ for all i, the desired result follows. Q.E.D.

Remark. It follows from Theorem 5.1 a) that every complete, locally m-convex algebra (resp. locally convex space) is a projective limit of Banach algebras (resp. Banach spaces). This projective limit satisfies LP I of [28, p.23], but not LP II and LP III.

---

(4) $\mathcal{LP}(A)$ is defined like the projective limit of the algebras A_i with respect to the homomorphisms π_{ij} (see [28]), but in general, condition LP II and LP III on p.24 of [28] are not satisfied here.

Among the various applications of Theorem 5.1, the following is probably the most important. (Other applications are found in Proposition 7.4, Theorem 8.4, Proposition 10.1, and Theorem 10.3).

Theorem 5.2. *Let A be a complete, locally m-convex algebra with m-base $\{U_i\}$, and let $x \in A$. Then*

a) *x is q.r. in A if and only if x_i is q.r. in $\overline{A}_i$ for all i.*

b) *A has a unit if and only if every $\overline{A}_i$ has a unit.*

c) *x is invertible in A if and only if x_i is invertible in $\overline{A}_i$ for all i.*

Proof. The "only if" parts of the theorem are all clear. Let us prove the "if" parts.

a) Suppose that x_i is q.r. in $\overline{A}_i$ for all i, and, for each i, let $(x_i)'$ be the quasi-inverse of x_i in $\overline{A}_i$. If $j \geq i$, then $[\overline{\pi}_{ij}((x_i)')] \circ x_i = [\overline{\pi}_{ij}((x_i)')] \circ [\overline{\pi}_{ij}(x_j)] = \overline{\pi}_{ij}((x_j)')\circ(x_j) = 0$, and therefore $\overline{\pi}_{ij}((x_j)') = (x_i)'$. Hence, by Theorem 5.1 b), there exists an element $y \in A$ such that $y_i = (x_i)'$ for all i. Now if $x \in A$, then $(x \circ y)_i = x_i \circ (x_i)' = 0$ for all i, and therefore $x \circ y = 0$. Similarly $y \circ x = 0$, and thus x is q.r.

b) For all i, let e(i) be the unit of $\overline{A}_i$. If $j \geq i$, and if $x_i \in A_i$, then $[\overline{\pi}_{ij}(e(j))]x_i = [\overline{\pi}_{ij}(e(j))][\overline{\pi}_{ij}(x_j)] = \overline{\pi}_{ij}(e(j)x_j) = \overline{\pi}_{ij}(x_j) = x_i$. Since A_i is dense in $\overline{A}_i$, it follows that $\overline{\pi}_{ij}(e(j))$ is a left unit for $\overline{A}_i$. Similarly, $\overline{\pi}_{ij}(e_j)$ is a right unit for $\overline{A}_i$, and hence $\overline{\pi}_{ij}(\overline{e}_j) = \overline{e}_i$. Hence, by Theorem 5.1 b), there is an $e \in A$ such that $e_i = e(i)$ for all i. Hence, if $x \in A$, then $(xe)_i = x_i e(i) = x_i$ for all i, and therefore $xe = x$; similarly $ex = x$, and therefore e is a unit for A.

c) If x_i is invertible in $\overline{A}_i$ for all i, then A has a unit by (b). The proof that x has an inverse in A now goes exactly like the proof of (a).

Q.E.D.

Remark. Theorem 5.2 (especially part (a)) is one of the most useful in this paper. Among other things, it will presently enable us to extend to locally m-convex algebras certain properties of Banach algebras which the latter have by virtue of being Q-algebras (these are defined in Appendix E).

(Complete, locally m-convex algebras need not be Q-algebras, as is shown by Example 3.1.)

Most of the remainder of this section will be devoted to deriving some of the consequences of Theorem 5.2.

Corollary 5.3. *Let A be a complete, locally m-convex algebra, with m-base* $\{U_i\}$, *and let* $x \in A$. *Then*

a) $\Sigma_A(x) = \bigcup_i \Sigma_{\overline{A}_i}(x_i)$

b) $\sigma_A(x) = \sup_i \sigma_{\overline{A}_i}(x_i) = \sup_i \lim_{n\to\infty} \sqrt[n]{p_i(x^n)}$.

Proof.

a) For $\lambda \neq 0$, $\lambda \in \Sigma_A(x) \leftrightarrow -\lambda^{-1}x$ is not q.r. in A $\leftrightarrow -\lambda^{-1}x_i$ is not q.r. in $\overline{A}_i$ for some i $\leftrightarrow \lambda \in \Sigma_{\overline{A}_i}(x_i)$ for some i. For $\lambda = 0$, $\lambda \in \Sigma_A(x) \leftrightarrow$ x is not invertible in A $\leftrightarrow$ x is not invertible in $\overline{A}_i$ for some i $\leftrightarrow \lambda \in \Sigma_{\overline{A}_i}(x_i)$ for some i.

b) The first equality follows from (a), and the second equality follows from the formula for the spectral radius in a Banach algebra [24, Lemma 1.1]. Q.E.D.

Theorem 5.4. *Let A be a commutative, complete locally m-convex algebra, and let* $x \in A$. *Then x is q.r. in A if and only if* $f(x) \neq -1$ *for all* $f \in \mathcal{M}(A)$.

Proof. If x is q.r. in A, then f(x) is q.r., and hence not -1, for all $f \in \mathcal{M}(A)$.

Suppose, on the other hand, that x is not q.r. in A, and let $\{U_i\}$ be any m-base for A. Then there exists an i such that x_i is not q.r. in $\overline{A}_i$. Since $\overline{A}_i$ is a commutative Banach algebra, we can therefore find an $f \in \mathcal{M}(\overline{A}_i)$ such that $f(x) = -1$. But then $f \circ \pi_i \in \mathcal{M}(A)$, and $f \circ \pi_i(x) = -1$.

Q.E.D.

Corollary 5.5. *Let A be a commutative, complete, locally m-convex algebra, and let* $x \in A$. *Then* $\mathcal{R}(A) = \{x \in A \mid f(x) = 0$ *for all* $f \in \mathcal{M}(A)\}$; *in particular* $\mathcal{R}(A)$ *is closed.*

Proof. Denoting the set on the right by K, we certainly have $\mathcal{R}(A) \in K$. If $x \in K$, then $f(x) \neq -1$ for all $f \in \mathcal{M}(A)$, whence x is q.r. Thus K is a q.r. ideal, and therefore $K \subset \mathcal{R}(A)$. Q.E.D.

Corollary 5.6. *Let A be a commutative, complete, locally m-convex algebra, and let* $x \in A$. *Then*

a) $\Sigma_A(x) \cup \{0\} = \{f(x) \mid f \in \mathcal{M}(A)\}$.

b) $\sigma_A(x) = \sup \{|f(x)| \mid f \in \mathcal{M}(A)\}$.

Proof.

a) For $\lambda \neq 0$, $\lambda \in \Sigma_A(x) \leftrightarrow -\lambda^{-1}x$ is not q.r. $\leftrightarrow f(-\lambda^{-1}x) = -1$ for some $f \in \mathcal{M}(A) \leftrightarrow f(x) = \lambda$ for some $f \in \mathcal{M}(A)$.

b) This follows immediately from (a).

Q.E.D.

Remark. $\sigma_A(x)$ may of course be $+\infty$. For the case where $\sigma_A(x) < \infty$ for all x, see Section 13.

Corollary 5.7. *Let A be a complete, locally m-convex algebra, and let x, y be commuting elements of A. Then*

a) $\sigma_A(x + y) \leq \sigma_A(x) + \sigma_A(y)$

b) $\sigma_A(xy) \leq \sigma_A(x)\sigma_A(y)$

c) $\sigma_A(x^n) = (\sigma_A(x))^n$.

Proof. Let B be a maximal commutative subalgebra of A which contains x and y. Clearly $\Sigma_B(z) = \Sigma_A(z)$ for all $z \in B$. Since B is closed in A, and therefore complete, the results follow by applying Corollary 5.6 b) to B.

Q.E.D.

Before making our next application of Theorem 5.1, we need half of the following

Lemma 5.8. *Let A be a locally m-convex algebra with m-base* $\{U_i\}$, *and let* $x \in A$. *Then* x_i *is q.r. in* $\bar{A}_i$ *for all i if and only if -x is not a unit modulo any proper, closed, regular one-sided ideal in A.*

Proof. If, for some i, x_i is not q.r. in $\bar{A}_i$, then there exists a regular, maximal, one-sided ideal J in $\bar{A}_i$ with unit $-x_i$. Since $\bar{A}_i$ is a Q-algebra, J is closed in $\bar{A}_i$. Hence $\pi_i^{-1}(J)$ is a proper, closed, regular, one-sided ideal in A modulo which -x is a unit.

Suppose now that -x is a unit modulo some proper, closed, regular one-sided ideal J in A. By Lemma 7.1, there exists an i such that $\pi_i(J)$

is not dense in A_i. Now x_i is a unit modulo $\pi_i(J)$ in A_i, and hence also modulo K, the closure of $\pi_i(J)$ in A_i. It follows, by Lemma 3.10, that $-x$ is a unit modulo the closure of K in $\overline{A}_i$, and hence x is not q.r. in $\overline{A}_i$.

Q.E.D.

We now have the following important theorem.

Theorem 5.9. *Let A be a (commutative) complete, locally m-convex algebra, and let $x \varepsilon A$. Then x is q.r. in A if and only if $-x$ is not a unit modulo any proper, closed, regular (maximal) one-sided ideal in A.*

Proof. The "only if" assertion is trivial. The commutative part of the theorem follows immediately from Theorem 5.4. Let us therefore prove the remainder of the "if" assertion.

Suppose that x is not q.r. in A, and pick any m-base $\{U_i\}$ for A. By Theorem 5.2, there exists an i such that x_i is not q.r. in $\overline{A}_i$. Hence, by Lemma 5.8, there exists a proper, closed, regular, one-sided ideal in A modulo which $-x$ is a unit.

Q.E.D.

Remark. For an example where the conclusion of Theorem 5.9 is false, see [15, p.159].

Some implications of Theorem 5.9 will be found in Section 7 and Section 9. The question of whether we can find a *maximal* ideal when A is *not commutative* is discussed in Section 9.

We now conclude this section with the following proposition, which shows that the part of Theorem 5.1 which deals with q.r. elements is also true in (not necessarily complete!) locally m-convex Q-algebras. This of course implies that most of the consequences of Theorem 5.1 are true for Q-algebras as well as complete ones. We never mention this fact explicitly; it is not very useful, since most of these consequences (such as Theorem 5.9) are trivial for Q-algebras.

Proposition 5.10. *If A is a locally m-convex Q-algebra with m-base $\{U_i\}$, and if $x \varepsilon A$, then x is q.r. in A if and only if x_i is q.r. in $\overline{A}_i$ for all i.*

Proof. To prove the non-trivial part of the proposition, suppose that x is not q.r. in A. Then there exists a regular, maximal, one-sided ideal J in A modulo which $-x$ is a unit, and J is closed since A is a Q-algebra. Hence, by Lemma 5.8, x_i is not q.r. in $\bar{A}_i$ for some i.

Q.E.D.

6. Multiplicative linear functionals and symmetry.

Lemma 6.1. Let A be an algebra, f a multiplicative linear functional on A, and $x \in A$. Then

a) $f(x) \in \Sigma_A(x)$.

b) If A is a commutative, complete, locally m-convex algebra, then $f(x) \in \{g(x) \mid g \in \mathcal{M}(A)\}$.

Proof.

a) Let $f(x) = \lambda$. If $\lambda = 0$, then certainly x cannot have an inverse, and thus $\lambda \in \Sigma_A(x)$. If $\lambda \neq 0$, then $-\lambda^{-1}x$ cannot be q.r.; for if it were, then $f(-\lambda^{-1}x) = -\lambda^{-1}f(x) = -1$ would be q.r. which it is not.

b) This follows from (a) and Corollary 5.6 a).

Q.E.D.

Turning to $\mathcal{M}(A)$, we have

Lemma 6.2.

a) $\mathcal{M}(A)$ is closed in A' with the weak topology $\sigma(A', A)$.

b) If A has a unit e, then 0 is an isolated element of A, or, equivalently, $\mathcal{M}^-(A)$ is closed in $\mathcal{M}(A)$.

c) If A is an $\mathcal{F}$-algebra (with unit), then $\mathcal{M}(A)$ (and $\mathcal{M}^-(A)$) is hemi-compact.(5)

Proof.

a) $\mathcal{M}(A) = \bigcap_{x,\, y \in A} \{f \in A' \mid f(xy) - f(x)f(y) = 0\}$. Since, for fixed x and y, the maps $f \longrightarrow f(x)$, $f \longrightarrow f(y)$ and $f \longrightarrow f(xy)$ are all

---

(5) According to Arens [1, p.486], a topological space X is hemi-compact if there exists a sequence $\{C_n\}_{n=1}^{\infty}$ of compact subsets of X whose union is X and such that every compact subset of X is contained in some C_i. Every topological space which is both σ-compact and locally compact is hemi-compact.

continuous on A', it follows that $\mathcal{M}(A)$ is the intersection of closed sets, and hence closed.

b) Clearly 0 is the only element of the open set $\{f \varepsilon \mathcal{M}(A) | \ |f(e)|<1\}$.

c) If A is an $\mathcal{F}$-space, with base $\{U_n\}_{n=1}^{\infty}$ for the neighborhoods of 0, then A', with the topology $\sigma(A', A)$, is the union of the compact sets U_n°. If C is any compact subset of A', then C° is a neighborhood of 0 in A by Proposition 4.2, a $\longrightarrow$ e; hence $C^{\circ} \supset U_n$ for some n, and therefore $C \subset C^{\circ\circ} \subset U_n^{\circ}$. Thus A' is hemicompact, and hence so is $\mathcal{M}(A)$ by (a).

If A has a unit, then $\mathcal{M}^{-}(A)$ is closed in $\mathcal{M}(A)$ by (b), and is therefore also hemicompact.

Q.E.D.

For further information about $\mathcal{M}(A)$, see the next section.

We now turn to symmetric *-algebras. Following the usual conventions, we make

<u>Definition 6.3.</u> A <u>star</u> <u>algebra</u> (or <u>*-algebra</u>) is an algebra with a mapping which assigns to every x ε A and x^{*} ε A such that, for x, y ε A,

a) $(x^{*})^{*} = x$

b) $(xy)^{*} = y^{*}x^{*}$

c) $(x + y)^{*} = x^{*} + y^{*}$

d) $(\lambda x)^{*} = \bar{\lambda} x^{*}$ for any scalar λ.

A <u>symmetric</u> <u>*-algebra</u> is a *-algebra with the property that xx^{*} is q.r. for every x ε A.

Lemma 6.4. <u>Let</u> A <u>be</u> <u>a</u> <u>symmetric</u> <u>*-algebra,</u> <u>and</u> <u>let</u> x ε A.

a) <u>If</u> $x = x^{*}$, <u>then</u> $\Sigma_A(x)$ <u>is</u> <u>real,</u> <u>and</u> <u>if</u> $x = -x^{*}$, <u>then</u> $\Sigma_A(x)$ <u>is</u> <u>imaginary.</u>

b) If f is a multiplicative linear functional on A, then $f(x^{*}) = \overline{f(x)}$.

Proof.

a) This follows from [18, Lemma 4.1].

b) By (a) and Lemma 6.1 a), $f(x^{*}) - f(x)$ is real and $f(x^{*}) + f(x)$ is imaginary. Thus $f(x^{*}) = \overline{f(x)}$.

Q.E.D.

From Lemma 6.4 b) we immediately have

Corollary 6.5. <u>If A is a symmetric *-algebra of complex-valued functions on a space T, then A is closed under conjugation (i.e., if $x \in A$, then the function x^*, defined by $x^*(t) = \overline{x(t)}$ for $t \in T$, is in A).</u>

Lemma 6.6. <u>Let A be a symmetric *-algebra, and B a commutative, semi-simple, and complete locally m-convex symmetric *-algebra. Then any homomorphism h from A into B is a *-homomorphism (i.e., $h(x^*) = [h(x)]^*$ for all $x \in A$).</u>

Proof. Let $x \in A$. Then, for all $f \in \mathcal{M}(A)$, $f(h(x^*) - [h(x)]^*) = [f \circ h](x^*) - f([h(x)]^*) = \overline{[f \circ h](x)} - \overline{f[h(x)]}$ (by Lemma 6.4 b)) = 0. Hence $h(x^*) - (h(x))^* = 0$ by Corollary 5.5.

Q.E.D.

Lemma 6.7. <u>If A is a commutative, complete, locally m-convex, symmetric *-algebra, then A^+, with the *-operation $(x, \lambda)^* = (x^*, \overline{\lambda})$, is also symmetric.</u>

Proof. Certainly A^+ is complete. If $f \in \mathcal{M}(A^+)$, and if $(x,\lambda) \in A^+$, then $f[(x,\lambda)(x,\lambda)^*] - [f(x) + \lambda]^2 \neq 1$. Hence, by Theorem 5.4, $(x,\lambda)(x,\lambda)^*$ is q.r.

Q.E.D.

We will conclude this section with the following proposition about the algebras of Corollary 6.5.

Proposition 6.8. <u>Let T be a topological space, $t \in T$, and A a separating subalgebra of $C_t(T)$ which is closed under conjugation. Then A is dense in $C_t(T)$ with the compact-open topology.</u>

Proof. We need only show that for every compact subset D of T which does not (resp. does) contain t, the algebra $A|D$ of all restrictions of elements of A to D is a dense subalgebra of C(D) (resp. $C_t(D)$) in the topology of uniform convergence on D. But this follows from the Stone-Weierstrass theorem [27, Corollary 2, p.238], since D is compact, $A|D$ is closed under conjugation and separating, and at no point of D (resp. only at t) does every element of $A|D$ vanish.

Q.E.D.

7. *Miscellaneous relations between* A *and* $\{A_i\}$. Throughout this section, A will denote a locally m-convex algebra, and $\{U_i\}$ will denote an m-base for A.

Lemma 7.1. *If* S *is a subset of* A, *then* S *is dense in* $\bigcap_i \pi_i^{-1}(\overline{\pi_i(S)})$, *where* $\overline{\pi_i(S)}$ *denotes the closure of* $\pi_i(S)$ *in* A_i.

Proof. Denote $\bigcap_i \pi_i^{-1}(\pi_i(S))$ by $\hat{S}$. We clearly need only show that $\pi_i(S)$ is dense in $\pi_i(\hat{S})$ for all i. But this is certainly so, since $\pi_i(\hat{S}) \subset \overline{\pi_i(S)}$ for all i.

Q.E.D.

In the following proposition, "ideal" stands for either a right, a left, or a two-sided ideal (consistently!).

Proposition 7.2. *Let* I *be a subset of* A, *and consider the following properties of* I:

a) $I = \pi_i^{-1}(J)$ *for some closed, regular, maximal ideal* J *in some* A_i.

b) I *is a closed, regular, maximal ideal in* A.

c) $I = \pi_i^{-1}(J)$ *for some closed, regular, maximal ideal* J *in some* $\overline{A}_i$.

d) $I = \pi_i^{-1}(J)$ *for some regular maximal ideal* J *in some* A_i.

e) I *is a regular, maximal ideal in* A.

Then the following implications hold:

$$\begin{array}{ccccc} a & \longrightarrow & b & \longrightarrow & c \\ & & \downarrow & & \\ & & d & & \\ & & \downarrow & & \\ & & e & & \end{array}$$

If A *is commutative, then* (a), (b), *and* (c) *are equivalent.*

Proof.

a $\longrightarrow$ b: Obvious.

b $\longrightarrow$ c: By Lemma 7.1, there exists an i such that K, the closure of $\pi_i(I)$ in A_i, is a proper subset of A_i. Since $\pi_i(I)$ is regular in A_i, it follows that K is a proper, closed, regular ideal in A_i. By Lemma B.12, there exists a regular, maximal ideal J in $\overline{A}_i$ such that $K \subset J \cap A_i$. Hence $I \subset \pi_i^{-1}(J)$, and since I is maximal, $I = \pi_i^{-1}(J)$.

b $\longrightarrow$ d: We obtain $K \subset A_i$ as in the proof of b $\longrightarrow$ c. By Zorn's lemma, we may embed K in a regular, maximal ideal J in A_i. Then $I \subset \pi_i^{-1}(J)$, and since I is maximal, $I = \pi_i^{-1}(J)$.

d $\longrightarrow$ e. Obvious.

If A is commutative, then so is every $\overline{A}_i$. Now if J is closed, regular, maximal in $\overline{A}_i$, then J is the kernel of a non-zero, continuous, multiplicative linear functional on A. Hence $J \cap A_i$ is the kernel of a non-zero continuous, multiplicative linear functional in A_i, and therefore $J \cap A_i$ is a closed, regular, maximal ideal in A_i. Hence c $\longrightarrow$ a, which is all we need to show.

Q.E.D.

Remark. For not necessarily commutative algebras where (a), (b), and (c) are equivalent, see Proposition 9.6.

We now apply what we know to the study of the radical. For convenience, let us introduce the following notation.

Notation: $C\mathcal{R}(A)$ denotes the intersection of the closed, regular, maximal, right ideals of A.

Remark. Observe that, for commutative A, it follows from Proposition 2.9 c) that $C\mathcal{R}(A) = \{x \in A \mid f(x) = 0 \text{ for all } f \in \mathcal{M}(A)\}$.

Proposition 7.3.

a) $\mathcal{R}(A) \subset \bigcap_i \pi_i^{-1}(\mathcal{R}(A_i))$; $\bigcap_i \pi_i^{-1}(\mathcal{R}(\overline{A}_i)) \subset C\mathcal{R}(A)$.

b) _If_ A _is_ _commutative_, _then_
$\mathcal{R}(A) \subset \bigcap_i \pi_i^{-1}(\mathcal{R}(A_i)) \subset \bigcap_i \pi_i^{-1}(\mathcal{R}(\overline{A}_i)) = C\mathcal{R}(A)$.

c) _If_ A _is_ _complete_, _then_
$\bigcap_i \pi_i^{-1}(\mathcal{R}(\overline{A}_i)) \subset \mathcal{R}(A) = \bigcap_i \pi_i^{-1}(\mathcal{R}(A_i)) \subset C\mathcal{R}(A)$.

d) _If_ A _is_ _commutative_ _and_ _complete_, _then_
$\mathcal{R}(A) = C\mathcal{R}(A) = \bigcap_i \pi_i^{-1}(\mathcal{R}(\overline{A}_i)) = \bigcap_i \pi_i^{-1}(\mathcal{R}(A_i))$.

Proof.

a) To prove the first inclusion we need only observe that, if B and C are algebras, and h a homomorphism from B onto C, then $h(\mathcal{R}(A)) \subset \mathcal{R}(B)$. The second inclusion follows from Proposition 7.2, b $\longrightarrow$ c.

b) The first inclusion and half of the equality follow from (a). The second inclusion follows from Proposition 7.2, c$\longrightarrow$a (commutative case!). Finally, $\mathcal{C}\mathcal{R}(A) \subset \bigcap_i \pi_i^{-1}(\mathcal{R}(\overline{A}_i))$ follows from Proposition 7.2, c $\longrightarrow$ b (commutative case!).

c) We need only prove that $\bigcap_i \pi_i^{-1}(\mathcal{R}(\overline{A}_i))$ and $\bigcap_i \pi_i^{-1}(\mathcal{R}(A_i))$ are both contained in $\mathcal{R}(A)$. Since both of them are two-sided ideals in A, it is only necessary to show that every element of these ideals is q.r. in A. But if x is such an element, then, for every i, x_i is q.r. in $\overline{A}_i$; hence x is q.r. in A by Theorem 5.2.

d) This follows from (b) and (c).

Q.E.D.

Remark. For not necessarily commutative algebras where the results of (b) and (d) are valid, see Proposition 9.6.

We leave the reader to draw the obvious conclusions concerning semi-simplicity. We also raise the following question to which we do not know the answer, even if A is a commutative $\mathcal{F}$-algebra: If A is a semi-simple, locally m-convex algebra, does there exist some m-base $\{U_i\}$ for A such that $\overline{A}_i$ is semi-simple for all i ?

Proposition 7.4. <u>If A is commutative and complete, and if every $\overline{A}_i$ is a semi-simple symmetric *-algebra, then A is a semi-simple symmetric *-algebra.</u>

Proof. If every $\overline{A}_i$ is semi-simple, then so is A, by Proposition 7.3 (b).

Let us now turn to symmetry. Let $x \varepsilon A$ and, using the notation developed at the beginning of Section 5, let $j \geq i$. Then $\overline{\pi}_{ij}((x_j)^*) = (x_i)^*$ by Lemma 6.6. Hence, by Theorem 5.1 (b), there is an element $z \varepsilon A$ such that $z_i = (x_i)^*$ for all i. Now let $x^* = z$. Let us verify that this is a proper *-operation: We have $((x + y)^*)_i = ((x + y)_i)^* = (x_i + y_i)^* = (x^*)_i + (y^*)_i$ for all i, and hence $(x + y)^* = x^* + y^*$; the other properties are verified similarly. To see that A is symmetric with this *-operation, observe that $(xx^*)_i = x_i(x^*)_i = x_i(x_i)^*$ which is q.r. in $\overline{A}_i$ by assumption; hence xx^* is q.r. in A by Theorem 5.2.

Q.E.D.

Let us now turn to the study of $\mathcal{M}(A)$, or, equivalently, the set of closed, two-sided maximal ideals M in A such that A/M is isomorphic to the algebra of complex numbers. (If A is commutative, then, by Proposition 2.9 c), this coincides with the set of all closed, regular, maximal ideals in A.) We begin with the following observations:

If $f \in \mathcal{M}(A) \cap U_i^o$, then f vanishes on N_i (see Notation 2.6), and we may therefore define a multiplicative linear functional f_i on A_i by "$f_i(x_i) = f(x)$". It is clear that $f_i \in \mathcal{M}(A_i)$. The mapping $f \to f_i$ from $\mathcal{M}(A) \cap U_i^o$ to $\mathcal{M}(A_i)$ is clearly 1-1; it is also onto, since for any $g \in \mathcal{M}(A_i)$, $g \circ \pi_i \in \mathcal{M}(A) \cap U_i^o$, and $g = (g \circ \pi_i)_i$. Let us now go on. If $f_i \in \mathcal{M}(A_i)$, then we may extend f_i uniquely to a functional $\overline{f}_i$ on $\overline{A}_i$, and clearly $\overline{f}_i \in \mathcal{M}(\overline{A}_i)$. The map $f_i \to \overline{f}_i$ from $\mathcal{M}(A_i)$ to $\mathcal{M}(\overline{A}_i)$ is clearly one-to-one and onto. We may summarize the connection between f, f_i, and $\overline{f}_i$ by the following equation:

(*) $$f(x) = f_i(x_i) = \overline{f}_i(x_i) \qquad \text{for all } x \in A.$$

The further relation between $\mathcal{M}(A)$ and the $\mathcal{M}(A_i)$'s may now be summarized in

Proposition 7.5.

a) $\mathcal{M}(A) = \bigcup_i [\mathcal{M}(A) \cap U_i^o]$; $\mathcal{M}^-(A) = \bigcup_i [\mathcal{M}^-(A) \cap U_i^o]$.

b) $\mathcal{M}(A) \cap U_i^o$ *is compact for all* i.

c) *The* 1-1 *mappings* $f \to f_i$ *from* $\mathcal{M}(A) \cap U_i^o$ *onto* $\mathcal{M}(A_i)$, *and* $f_i \to \overline{f}_i$ *from* $\mathcal{M}(A_i)$ *onto* $\mathcal{M}(\overline{A}_i)$, *are homeomorphisms. When restricted to* $\mathcal{M}^-(A) \cap U_i^o$ (*resp.* $\mathcal{M}^-(A_i)$), *they are homeomorphisms from* $\mathcal{M}^-(A) \cap U_i^o$ *onto* $\mathcal{M}^-(A_i)$ (*resp. from* $\mathcal{M}^-(A_i)$ *onto* $\mathcal{M}^-(\overline{A}_i)$).

Proof.

a) If $f \in \mathcal{M}(A)$, then $f(U_{i_o})$ is bounded for some i_o. If $f \notin (U_{i_o})^o$, then $|f(x)| > 1$ for some $x \in U_{i_o}$. But then $\{|f(x^n)|\}_{n=1}^{\infty} = \{|f(x)|^n\}_{n=1}^{\infty}$ is not bounded; since $x^n \in U_{i_o}$ for all n, this is a contradiction. The second assertion follows from the first.

b) This follows from Lemma 6.2 a), and the fact that the polar of any neighborhood of the origin (such as U_i^o) is compact.

c) That $f \to f_i$ is a homeomorphism follows directly from the definition of the topologies on $\mathcal{M}(A)$ and $\mathcal{M}(A_i)$. That $f_i \to \overline{f}_i$ is a homeomorphism is well known. (Observe that the homeomorphic sets $\mathcal{M}(A) \cap U_i^o$, $\mathcal{M}(A_i)$, and $\mathcal{M}(\overline{A}_i)$ are all compact.) The assertion about the induced homeomorphisms is obvious.

Q.E.D.

It is well known that, if T is a compact Hausdorff space, and if $A = C(T)$ with the "sup" norm, then A is a Banach algebra, and $\mathcal{M}^-(A)$ is canonically isomorphic to T. If T is not compact, then this situation becomes a little more difficult. The following example shows what happens, and illustrates some of the concepts and theorems, in particular Lemma 2.5 and Proposition 7.5, which we have developed.

Example 7.6. Let T be a completely regular topological space, and let $\{D_i\}$ be a collection of compact subsets of T whose union in T, and such that every finite union of D_i's is contained in a D_i. Let $A = C(T)$, with the topology of uniform convergence on the D_i's. Then

a) If $U_i = \{x \in A \mid |x(t)| \leq 1 \text{ whenever } t \in D_i\}$, then $\{U_i\}$ is an m-base for A.

b) $N_i = \{x \in A \mid x(T) = 0 \text{ whenever } t \in D_i\}$, (see Notation 2.6).

c) Let $\psi_i\colon C(T) \to C(D_i)$ be defined by $\psi_i(x) = x|D_i$. Then ψ_i is onto $C(D_i)$, and has kernel N_i. Hence ψ_i establishes an algebraic isomorphism ψ_i' between A_i and $C(D_i)$. If $C(D_i)$ is normed with the "sup.norm", which we shall henceforth assume, then ψ_i' is norm preserving.

d) $A_i = \overline{A}_i$.

e) The mapping $u_i\colon D_i \to \mathcal{M}(A)$, defined by "$[u_i(t)]\,(x) = x(t)$", is a homeomorphism of D_i onto $\mathcal{M}^-(A) \cap U_i^o$.

f) The mapping $u\colon T \to \mathcal{M}^-(A)$, defined by "$[u(t)]\,(x) = x(t)$", is a homeomorphism of T onto $\mathcal{M}^-(A)$.

Proof. Assertions (a) and (b) are obvious. The only part of (c) which requires proof is that ψ_i is onto $C(D_i)$; this follows from [27, Th.13].

Assertion (d) is clear. Assertion (e) follows from Proposition 6.4 c), assertion (c) above, and the well known fact that $\mathcal{M}(C(D_i))$ and D_i are canonically homeomorphic. To prove (f), we first observe that u is onto by (e) and Proposition 7.5 (a), and that u is 1-1 since C(T) is separating on T. Finally, u is a homeomorphism, since the topology on a completely regular space T is the weakest topology for which all the elements of C(T) remain continuous.

Q.E.D.

Our next lemma deals with Cauchy sequences. (Parts (a) and (b) are also true for directed sets instead of sequences.)

Lemma 7.7.

a) *If* $\{x_n\}_{n=1}^{\infty}$ *is a sequence of elements of* A, *then* $\{x_n\}_{n=1}^{\infty}$ *is Cauchy in* A *if and only if* $\{(x_n)_i\}_{n=1}^{\infty}$ *is Cauchy in* A_i *for each* i.

b) *If* A *is complete, if* $\{x_n\}_{n=1}^{\infty}$ *is a sequence of elements of* A *which is a bounded subset of* A, *and if the series* $\sum_{n=1}^{\infty} \lambda_n$ *of scalars converges absolutely, then the series* $\sum_{n=1}^{\infty} \lambda_n x_n$ *converges in* A.

c) *If* A *is complete, if* $\{\lambda_n\}_{n=1}^{\infty}$ *is the sequence of coefficients of a series representing an entire function, and if* $x \in A$, *then* $\sum_{n=1}^{\infty} \lambda_n x^n$ *converges to an element of* A. *The map* $x \longrightarrow \sum_{n=1}^{\infty} \lambda_n x^n$ *is continuous.*

Proof.

a) This follows immediately from the definition.

b) This follows from (a), and the fact that, for each i, the series $\sum_{n=1}^{\infty} \lambda_n (x_n)_i$ converges absolutely.

c) This is proved in [2, Th. 3]. The convergence follows easily from (a), since for every i, the series $\sum_{n=1}^{\infty} \lambda_n (x^n)_i = \sum_{n=1}^{\infty} \lambda_n (x_i)^n$ converges absolutely. Q.E.D.

Our final subject in this section is the case of a locally m-convex algebra $\hat{A}$, with subalgebra A. Let $\{\hat{U}_i\}$ be an m-base for $\hat{A}$, and equip the associated objects with hats (like $\hat{\pi}_i$, $\hat{A}_i$, $\hat{N}_i$, and $\hat{p}_i$). Let $U_i = \hat{U}_i \cap A$; then clearly $\{U_i\}$ is an m-base for A in the relative topology.

Lemma 7.8. <u>For each i, there is an isometry ϕ_i from A_i onto $\hat{\pi}_i(A)$ (considered as a subalgebra of the normed algebra $\hat{A}_i = \hat{\pi}_i(\hat{A})$) such that $\phi_i[\pi_i(x)] = \hat{\pi}_i(x)$ for all $x \in A$.</u>

Proof. If x and y are elements of A such that $\pi_i(x) = \pi_i(y)$, then $x - y \in N_i \subset \hat{N}_i$, and thus $\hat{\pi}_i(x) = \hat{\pi}_i(y)$. We may therefore define ϕ_i: $A_i \longrightarrow \hat{A}_i$ by $\phi_i[\pi_i(x)] = \hat{\pi}_i(x)$. Then obviously $\phi_i(A_i) = \hat{\pi}_i(A)$, and ϕ_i is an isometry because $p_i(x) = \hat{p}_i(x)$ for all $x \in A$.

Q.E.D.

8. <u>Functional representation theory.</u>

Proposition 8.1. <u>If A is a commutative, semi-simple, and complete locally m-convex algebra, then A is algebraically isomorphic to a separating subalgebra $\tilde{A}$ of $C_o(\mathcal{M}(A))$. Any element of $\tilde{A}$ which is never -1 has its quasi-inverse in $\tilde{A}$. If A is a symmetric *-algebra, then $\tilde{A}$ is closed under conjugation.</u>

Proof. Let ϕ: $A \longrightarrow C_o(\mathcal{M}(A))$ be defined by $[\phi(x)](f) = f(x)$, and let $\tilde{A} = \phi(A)$, Now ϕ is clearly a homomorphism; since the kernel of ϕ is 0 by Corollary 5.5, ϕ is an algebraic isomorphism. That $\tilde{A}$ is separating follows from its definition.

If $[\phi(x)](f) \neq -1$ for all $f \in \mathcal{M}(A)$, then x is q.r. in A by Theorem 5.4; if x' is the quasi-inverse of x in A, the $\phi(x')$ is the quasi-inverse of $\phi(x)$ in $\phi(A) = \tilde{A}$.

If A is a symmetric *-algebra, then $\tilde{A}$ is closed under conjugation by Corollary 6.5.

Q.E.D.

Remark. Observe that we indirectly used the Gelfand-Mazur theorem for normed algebras, but that we did not use its extension (Proposition 2.9) to locally m-convex algebras.

Remark. In the same way as above, A is isomorphic to a subalgebra $\widehat{A}$ of $C(\mathcal{M}^-(A))$. If A has a unit, then the identity function is in $\widehat{A}$, and every element of $\widehat{A}$ which is never 0 has its inverse in $\widehat{A}$.

Whenever A is thus algebraically isomorphic to a subalgebra of $C_o(\mathcal{M}(A))$, we will henceforth freely identify A (algebraically!) with this subalgebra.

The following observations will be helpful later on: Let A be a locally m-convex algebra, and let E be an equicontinuous subset of $\mathcal{M}(A)$. Then $E \subset U^o \cap \mathcal{M}(A)$ for some neighborhood U of 0 in A; since U^o is compact in A' with the topology $\sigma(A', A)$, and $\mathcal{M}(A)$ is closed in A' with this topology, it follows that $U^o \cap \mathcal{M}(A)$ is a compact subset of $\mathcal{M}(A)$. Suppose now that A is also commutative, complete, and semi-simple (and therefore identified with a subalgebra of $C_o(\mathcal{M}(A))$). Then, in view of the above, the following topologies on A are identical:

i) The topology of uniform convergence on the equicontinuous subsets of $\mathcal{M}(A)$.

ii) The topology of uniform convergence on the compact, equicontinuous subsets of $\mathcal{M}(A)$.

For convenience, we will denote this topology by "$\tau_o(A)$" throughout the remainder of this section. If A satisfies condition (h) of Proposition 4.2, (and particularly if A is an $\mathcal{F}$-algebra), then clearly $\tau_o(A)$ is the topology of uniform convergence on all compact subsets of $\mathcal{M}(A)$ (i.e., the compact-open topology).

Proposition 8.2. *Let A be a commutative, semi-simple, and complete locally m-convex algebra, with topology* τ. *Then*

a) τ *is finer than* $\tau_o(A)$.

b) *If A is a symmetric *-algebra, then A is dense in* $C_o(\mathcal{M}(A))$ *with the compact-open topology.*

Proof.

a) This follows from the definition of $\tau_o(A)$.

b) This follows from Proposition 8.1 and Proposition 6.8.

Q.E.D.

We now introduce a new concept.

Definition 8.3. A commutative, semi-simple, and complete locally m-convex algebra A is full if $A = C_o(\mathcal{M}(A))$ (algebraically!).

The theorem and proposition below will give some answers to the following two questions.

1) Under what conditions is an algebra full?

2) What can be said about the topology of a full algebra?

Question 1 is answered in Theorem 8.4 below; Conditions (a) and (e) of that theorem generalize similar conditions for Banach algebras, while Condition (b) reduces the question to the case of Banach algebras. As for Question 2, we conjecture that every full algebra A carries the topology $\tau_o(A)$. While we are unable to prove this in general, we do shed some light on the matter in Theorem 8.4 and Proposition 8.5. (Observe that a full algebra A need not always carry the topology of uniform convergence on all compact subsets of $\mathcal{M}(A)$, as is shown by Example 3.8. See, however, Proposition 8.5.)

Theorem 8.4. If A is a commutative and complete locally m-convex algebra with topology τ, then the following assertions are all equivalent.

a) A is a *-algebra, and there exists an m-base $\{U_i\}$ for A such that $p_i(xx^*) = (p_i(x))^2$ for all $x \varepsilon A$ and all i.

b) There exists an m-base $\{U_i\}$ for A such that every $\overline{A}_i$ is full.

c) A is full, and there exists an m-base $\{U_i\}$ for A such that every $\overline{A}_i$ is semi-simple.

d) A is full, and $\tau = \tau_o(A)$.

e) A is a semi-simple symmetric $*$-algebra, and $\tau = \tau_o(A)$.

Proof.

(a) $\rightarrow$ (b). Consider a fixed i. Our condition on p_i implies that $p_i(x) = p_i(x^*)$ for all $x \varepsilon A$, and hence $(y - z)^* \varepsilon N_i$ whenever $y - z \varepsilon N_i$. We may therefore make A_i into a $*$-algebra by setting $(x_i)^* = (x^*)_i$. Also $p_i(x_i(x_i)^*) = (p_i(x_i))^2$ for every $x_i \varepsilon A_i$. We may therefore extend the $*$-operation to $\overline{A}_i$ by continuity, and then the norm on $\overline{A}_i$ also satisfies the last equation for every element of $\overline{A}_i$. Hence $\overline{A}_i$ is a commutative C^*-algebra, which is well known to be full.

(b) $\rightarrow$ (c). Since every $\overline{A}_i$ is semi-simple, $\overline{A}_i$ is also semi-simple by Proposition 7.3 (d). We may therefore identify A with a subalgebra of $C_o(\mathcal{M}(A))$, and we must show that $A = C_o(\mathcal{M}(A))$. Let $z \varepsilon C_o(\mathcal{M}(A))$, and for each i, let $z(i) = z|\mathcal{M}(A) \cap U_i^o$. We know that $\mathcal{M}(A) \cap U_i^o$ and $\mathcal{M}(\overline{A}_i)$ are "naturally" homeomorphic (see Proposition 7.5); let x(i) be elements of $C_o(\mathcal{M}(\overline{A}_i))$ which corresponds to z(i) by means of this homeomorphism. Since $\overline{A}_i$ is full, we have $x(i) \varepsilon \overline{A}_i$. Using the notation of Theorem 5.1, it is now easily verified that $\overline{\pi}_{ij}(x(j)) = x(i)$ whenever $j \geq i$. By Theorem 5.1 (b), we therefore have an element $x \varepsilon A$ such that $x_i = x(i)$ for all i. By tracing the various identifications, we see that x = z. Hence every element of $C_o(\mathcal{M}(A))$ is in A, and therefore A is full.

The existence of an m-base $\{U_i\}$ such that every A_i is semi-simple follows directly from (b).

(c) $\rightarrow$ (b). Assume (c); we will show that every $\overline{A}_i$ is full. Pick a fixed i. Since $\overline{A}_i$ is semi-simple, we may identify $\overline{A}_i$ with a subalgebra of $C_o(\mathcal{M}(\overline{A}_i))$, and we must show that every z in $C_o(\mathcal{M}(\overline{A}_i))$ is also in $\overline{A}_i$. Now $\mathcal{M}(\overline{A}_i)$ is naturally homeomorphic to $\mathcal{M}(A) \cap U_i^o$ (see Proposition 7.5); let $y \varepsilon C_o(\mathcal{M}(A) \cap U_i^o)$ correspond to z by means of this homeomorphism. Since $\mathcal{M}(A)$ is completely regular, and since $\mathcal{M}(A) \cap U_i^o$ is compact, there exists an $x \varepsilon C_o(\mathcal{M}(A))$ such that $x|\mathcal{M}(A) \cap U_i^o = y$ (see [27, Theorem 13]). It follows from our construction that $\overline{f}_i(x_i) = \overline{f}_i(z)$ for all $\overline{f}_i \varepsilon \mathcal{M}(\overline{A}_i)$, and therefore $z = x_i \varepsilon \overline{A}_i$.

(b) $\rightarrow$ (d). Assume (b). We have already shown in the proof of (b) $\rightarrow$ (c) that this implies that A is full; it therefore remains to show that $\tau = \tau_o(A)$. Since τ is finer than $\tau_o(A)$ by Proposition 8.2 (a), we need only show that every U_i contains a $\tau_o(A)$-neighborhood of 0 in A. Since $\overline{A}_i$ is full, we know that it must carry topology of uniform convergence on $\mathcal{M}(\overline{A}_i)$. Consequently there exists a scalar λ_i such that, for every $z \varepsilon \overline{A}_i$, $p_i(z) \leq \lambda_i \sup_{\overline{f}_i \varepsilon \mathcal{M}(\overline{A}_i)} |\overline{f}_i(z)|$. Hence, for all $x \varepsilon A$,

$$p_i(x) = \tilde{p}_i(x_i) \leq \lambda_i \sup_{\bar{f}_i \varepsilon \mathcal{M}(\bar{A}_i)} |\bar{f}_i(x_i)| = \lambda_i \sup_{f \varepsilon \mathcal{M}(A) \cap U_i^o} |f(x)|.$$

Since $\mathcal{M}(A) \cap U_i^o$ is an equicontinuous subset of $\mathcal{M}(A)$, this proves our assertion.

(d) $\leftrightarrow$ (e). That (d) $\rightarrow$ (e) is obvious, and that (e) $\rightarrow$ (d) follows from Proposition 8.2 (b).

(e) $\rightarrow$ (a). By Lemma 6.4 (b), $f(x^*) = \overline{f(x)}$ for every $x \varepsilon A$ and $f \varepsilon \mathcal{M}(A)$. If $\{E_i\}$ is the collection of equicontinuous subsets of $\mathcal{M}(A)$, and if $U_i = \{x \varepsilon A |\ |f(x)| \leq 1$ for all $f \varepsilon E_i\}$, then $\{U_i\}$ is by definition an m-base for A. But then $p_i(x) = \sup_{f \varepsilon E_i} |f(x)|$ for every $x \varepsilon A$ and every i, and hence every p_i clearly satisfies the required equation $p_i(xx^*) = (p_i(x))^2$.

Q.E.D.

Remark. Observe that the equivalence oı (c) and (d) implies that the following two conjectures are equivalent:

i) If A is a full algebra with topology τ, then $\tau = \tau_o(A)$.

ii) If A is a full algebra, then there exists an m-base $\{U_i\}$ for A such that every $\bar{A}_i$ is semi-simple.

Proposition 8.5. <u>Let A be a full $\mathcal{F}$-algebra with topology τ, and suppose that $\mathcal{M}(A)$ is a k-space (see Appendix D). Then</u> $\tau = \tau_o(A)$ (= <u>the topology of uniform convergence on the compact subsets of $\mathcal{M}(A)$, as observed before Proposition</u> 8.2).

Proof. By Lemma 6.2 (c), $\mathcal{M}(A)$ is hemi-compact, and therefore $C(\mathcal{M}(A))$ is metrizable in the topology of uniform convergence on compact subsets of $\mathcal{M}(A)$. Since $\mathcal{M}(A)$ is a k-space, $C(\mathcal{M}(A))$ is also complete in this topology (Lemma D.5) and hence an $\mathcal{F}$-space. Thus the same is true of the closed subspace $C_o(\mathcal{M}(A))$. The result now follos from Proposition 8.2 (a), and the closed graph theorem [4, Th. 5, pg. 41]. Q.E.D.

Remark. The author does not know whether $\mathcal{M}(A)$ is <u>always</u> a k-space when A is a full $\mathcal{F}$-algebra.

Let us conclude this section with an application of Theorem 8.4, (b) $\rightarrow$ (d).

Corollary 8.6. *If $\{B_\mu\}$ is a collection of commutative C^*-algebras, and if A is a closed subalgebra of the cartesian product $\prod_\mu B_\mu$, then A is isomorphic to the algebra of all functions vanishing at a fixed point on a completely regular space T, in the topology of uniform convergence on some collection $\mathcal{S}$ of compact subsets of T. If $\{B_\mu\}$ is a countable collection, then T is hemi-compact, and $\mathcal{S}$ is the collection of all compact subsets.*

Proof. If we take the natural m-base for A, then each $\overline{A}_i$ is isomorphic to a closed subalgebra of a finite cartesian product of C^*-algebras, and hence each $\overline{A}_i$ is itself a C^*-algebra. Hence each $\overline{A}_i$ is full, and the first assertion follows from Theorem 8.4 (b) $\longrightarrow$ (d). The second assertion now follows from Lemma 6.2 (c), the fact that A is now an $\mathcal{F}$-space, and the fact that $\tau_0(A)$ is the topology of uniform convergence on *all* compact subsets of $\mathcal{M}(A)$ if A is an $\mathcal{F}$-space.

9. *Closed regular maximal one sided ideals.* We begin by proving two lemmas about closed, regular, maximal one sided ideals in arbitrary topological algebras.

Lemma 9.1. *Let A be a topological algebra. Then $\bigcap$ (closed, regular, maximal right ideals in A) = $\bigcap$ (primitive ideals associated with closed, regular, maximal, right ideals).*

Proof. Since every closed, regular, maximal right ideal contains the primitive ideal associated with it, the first set clearly contains the second. Suppose, further, that $P = \{x \in A \mid Ax \subset I\}$ is the primitive ideal associated with the closed, regular, maximal right ideal I. Then $P = \bigcap_{y \in A} \{x \in A \mid yx \in I\}$, and hence P is the intersection of the closed, regular, maximal, right ideals containing P. Therefore the second set in the lemma contains the first. Q.E.D.

Proposition 9.2. *Let A be a topological algebra, and consider the following properties of A:*

a) *If $-x$ is not a unit modulo any closed, regular, maximal, one sided ideal in A, then x is q.r.*

b) $\mathcal{R}(A) = \bigcap$ *(closed, regular, maximal, right ideals in A).*

c) *If I is a two sided ideal in A, then* $\mathcal{R}(I) = \bigcap$ *(closed, regular, maximal right ideals in I).*

d) *Every two sided ideal in A which is not a radical ring has a closed, regular maximal right ideal.*

e) $\mathcal{R}(A)$ *is closed.*

Let b', c', d' (*resp.* b", c", d") *be the assertions which we get from* b, c, *and* d, *if "right" is replaced by "left"* (*resp. "one-sided"*). *Then*

$$\begin{array}{ccccccccc} & & b & \longleftrightarrow & c & \longrightarrow & d & & \\ & & \downarrow & & \downarrow & & \downarrow & & \\ a & \longrightarrow & b'' & \longleftrightarrow & c'' & \longrightarrow & d'' & \longrightarrow & e \\ & & \uparrow & & \uparrow & & \uparrow & & \\ & & b' & \longleftrightarrow & c' & \longrightarrow & d' & & \end{array}$$

Proof. The vertical implications are all obvious. The implications in the top and bottom rows are all similar to the corresponding implications in the middle row; we will therefore only consider the latter. Now c" ⟶ b" and c" ⟶ d" are obvious. Let us prove the remaining assertions.

a ⟶ b": By Lemma 9.1, $\bigcap$ (closed, regular, maximal one-sided ideals) is a two sided ideal in A, and (a) implies that every element of this ideal is q.r. Hence this ideal is contained in $\mathcal{R}(A)$; the other inclusion is obvious.

b" ⟶ c": We need only observe that, if (b") is assumed, then $\mathcal{R}(I) \subset \bigcap$ (closed, regular, maximal one-sided ideals in I) $\subset I \cap \bigcap$ (closed, regular, maximal onesided ideals in A) = (by (b")) $I \cap \mathcal{R}(A) = \mathcal{R}(I)$.

d" ⟶ e: Following a suggestion of Kaplansky, we use the method of [19, Theorem 1]. Let R be the radical of A, with closure $\overline{R}$. Then $\mathcal{R}(\overline{R}) = R$, and we must show that $\overline{R}$ is a radical ring. If it were not, then there would exist a closed regular maximal one sided ideal J in $\overline{R}$. But then $R \subset J$, with R dense in $\overline{R}$ and J closed in $\overline{R}$, which is impossible.

Q.E.D.

Let us now turn to locally m-convex algebras A. We know from Theorem 5.4 that in this case all the properties of Proposition 9.2 are satisfied if A is commutative and complete. If only completeness is assumed, then we are unable even to prove property (e) of Proposition 9.2. However, under certain additional assumptions (one of which generalizes commutativity) we can show that property (b"), or even all the properties, of Proposition 9.2 hold. This will be the object of the remainder of this section.

Throughout the rest of this section, we assume that A is a locally m-convex algebra.

Theorem 9.3. <u>If A is normal (see Definition B.1) and complete, then property (a) of Proposition 9.2 holds.</u>

Proof. If x is not q.r., then, by Theorem 5.9, -x is a unit modulo some closed, regular, one-sided ideal in A. By normality, we can extend this ideal to a closed, regular, maximal one-sided ideal I in A, and -x will of course also be a unit modulo I.

Q.E.D.

Conditions for normality which apply to arbitrary topological algebras are given in Proposition B.5. In our case, we may sometimes apply these conditions indirectly through the following proposition.

Proposition 9.4. <u>If $\{U_i\}$ is an m-base for A such that every A_i is normal, then A is normal.</u>

Proof. Let I be a closed, regular, right ideal in A. By Lemma 7.1, there exists a i_0 such that $\pi_{i_0}(I)$ is not dense in A_{i_0}. Hence, by normality of A_{i_0}, there exists a closed, regular, maximal right ideal J in A_{i_0} which contains I. But then $\pi_{i_0}^{-1}(J)$ is a closed, regular, maximal right ideal in A which contains I, which proves the proposition.

Q.E.D.

We now make the following definition.

<u>Definition 9.5.</u> An m-base $\{U_i\}$ for A is <u>snug</u> if every A_i is snug in $\overline{A}_i$ (see Definition B.2).

Remark. If A is commutative, then it is easily seen (using Proposition 2.9 (c)) that every m-base for A is snug. This is a special case of Theorem 9.8 below.

Proposition 9.6.

a) *If A has a snug m-base, then A is normal.*

b) *Proposition 7.2 and Proposition 7.3 remain true if every requirement that A be commutative is replaced by the (weaker) requirement that $\{U_i\}$ be snug.*

Proof.

a) This follows from Proposition B.5(c) and Proposition 9.4.

b) The proof of Proposition 7.3 shows that it is sufficient to prove this for Proposition 7.2. We must therefore show that, if $\{U_i\}$ is snug, then condition (c) of Proposition 7.2 implies condition (a) of Proposition 7.2. But this follows from the definition (Definition 9.5) of a snug m-base.

Q.E.D.

Corollary 9.7. *If A has a snug m-base and is complete, then A has all the properties of Proposition 9.2.*

Proof. A has property (a) by Proposition 9.6 (a) and Proposition 6.3. A has properties (b) and (b') by Proposition 9.6 (b).

Q.E.D.

Sufficient conditions for a basis to be snug can be derived from Theorem B.7 and Proposition B.8. Besides the obvious conditions (in particular the condition that each A_i be equal to $\overline{A}_i$) we can derive the following

Theorem 9.8. *Each of the following conditions is sufficient for A to have a snug m-base (in (a) and (b) every m-base is snug).*

a) *A is completely continuous (Definition B.3).*

b) *A satisfies a polynomial identity.*

c) *A is a subalgebra of a cartesian product of completely continuous normed algebras.*

d) *A is a subalgebra of a cartesian product of normed algebras, each of which satisfies a polynomial identity.*

Proof. In proving (a) and (b), let $\{U_i\}$ be any m-base for A. By Theorem B.7, we need only show that every $\overline{A}_i$ is factor-finite. For (b) this follows from Lemma 9.9 below and Proposition B.8. For (a), we show it as follows: Let J be a closed, regular, maximal, right ideal in $\overline{A}_i$. To show that $\overline{A}_i/J$ is finite dimensional, we need only show that $A_i/(J \cap A_i)$ is finite dimensional, since $A_i/(J \cap A_i)$ is algebraically isomorphic to $(A_i + J)/J$ which is dense in $\bar{A}_i/J$. Let $I = \pi_i^{-1}(J \cap A_i)$; then I is a closed, regular, right ideal in A, and hence (by Lemma B.16) A/I is finite dimensional. Hence $A_i/(J \cap A_i)$ $(= \pi_i(A)/\pi_i(I))$ also is finite dimensional.

(c) and (d): Let A be a subalgebra of $\prod_{\mu \varepsilon M} A^{\mu}$, and let S^{μ} be the unit sphere of A. For each finite subset i of M, let $U_i = A \cap [(\prod_{\mu \varepsilon i} S^{\mu}) \times (\prod_{\mu \varepsilon M-i} A^{\mu})]$. Then $\{U_i\}$ is clearly an m-base for A, and A_i is a subalgebra of $\prod_{\mu \varepsilon i} S^{\mu}$ for all i. By Proposition B.9, it follows that every $\bar{A}_i$ is completely continuous (for case (c)) or satisfies a polynomial identity (for (d)). In either case, every $\bar{A}_i$ is factor-finite, by Proposition B.8, and hence every A_i is snug in $\bar{A}_i$ by Proposition B.7.

Q.E.D.

We conclude this section with the lemma promised in the proof of Theorem 9.8.

Lemma 9.9. Let A have m-base $\{U_i\}$. Then A satisfies the polynomial identity p if and only if every $\bar{A}_i$ satisfies p.

Proof. Let us first observe that every homomorphic image, as well as the completion, of a topological algebra which satisfies p also satisfies p. Hence, if A satisfies p, so does every A_i, and hence every $\bar{A}_i$.

Suppose, conversely, that every $\bar{A}_i$ satisfies p. If $x \varepsilon A$, then $\pi_i(p(x)) = p(\pi_i(x)) = 0$ for every i, and hence $p(x) = 0$.

Q.E.D.

10. Some theorems of Gelfand. In this section we extend Satz 20 and Satz 18 of Gelfand [12] to locally m-convex algebras.

We begin with Satz 20. While Gelfand assumes commutativity throughout [12], he clearly does not need it for his Satz 20. Observe that, even in the non-commutative case, the element y of Satz 20 of [12] commutes with x. We now extend Satz 20 as follows:

Theorem 10.1. Let A be a (commutative, semi-simple) complete locally m-convex algebra with unit e, let $x \varepsilon A$, and let w be a function of a complex variable which is analytic in some closed domain $D \supset \Sigma_A(x)$. Then there exists an (unique) element $y \varepsilon A$ commuting with x such that

$$f(y) = w[f(x)] \quad \text{for all} \quad f \varepsilon \mathcal{M}(A)$$

Proof. If such a y exists, then it is certainly unique for commutative, semi-simple A (Corollary 5.5).

Let us show that such a y exists. Let $\{U_i\}$ be an m-base for A, and pick an i. By [12, Satz 19], if Γ_i is any rectifiable Jordan curve contained in the domain of regularity of w which "surrounds" $\Sigma_{\overline{A}_i}(x_i)$ and does not intersect $\Sigma_{\overline{A}_i}(x_i)$, then the integral $\int_{\Gamma_i}(\lambda e_i - x_i)^{-1}w(\lambda)d\lambda$ converges to an element y(i) of $\overline{A}_i$ which is independent of the particular Γ_i chosen according to the above rules. We will now use Theorem 5.1 and the notation developed preceding it. If $j \geq i$, then $\overline{\pi}_{ij}(y(j)) = \overline{\pi}_{ij}\int_{\Gamma_j}(\lambda e_j - x_j)^{-1}w(\lambda)d\lambda = \int_{\Gamma_j}(\lambda e_i - x_i)^{-1}w(\lambda)d\lambda = \int_{\Gamma_i}(\lambda e_i - x_i)^{-1}w(\lambda)d\lambda$ (since $\Sigma_{\overline{A}_i}(x_i) \supset \Sigma_{\overline{A}_j}(x_j)) = y(i)$. By Theorem 5.1 (b), there exists a $y \varepsilon A$ such that $y_i = y(i)$ for all i. Now let $f \varepsilon \mathcal{M}(A)$. Then, by Proposition 7.5(a), there exists an i_o and an $f_{i_o} \varepsilon \mathcal{M}(\overline{A}_{i_o})$ such that $f(x) = f_{i_o}(x_{i_o})$ for all $x \varepsilon A$. Now by Satz 20 of [12], $f_{i_o}(y_{i_o}) = w[f_{i_o}(x_{i_o})]$; hence $f(y) = w[f(x)]$.

That x and y commute follows from the fact that x_i and y_i commute for all i.

Q.E.D.

We now turn to the generalization of Satz 18 of [12] to locally m-convex algebras. In Lemma 10.2 we will establish some preliminary facts about idempotents (in a slightly stronger form than necessary), in Theorem 10.3 we will generalize the crucial argument in the proof of Satz 18 of [12], and in Corollary 10.4 we will finally generalize Satz 18 of [12] itself.

Lemma 10.2. If c and d are distinct commuting idempotents in a complete, locally m-convex algebra A, then $\{0\} \neq \Sigma_A(c - d) \subset \{0, 1, -1\}$, and consequently $\sigma_A(c - d) = 1$.

Proof. By considering a maximal, commutative subalgebra of A containing c and d, we may restrict ourselves to the case where A is commutative. If $f \varepsilon \mathcal{M}(A)$, then f(c) and f(d) must be 0 or 1, and therefore f(c - d) must be 0, 1, or -1; by Corollary 5.6, it follows that $\Sigma_A(c - d) \subset \{0, 1, -1\}$.

If $\Sigma_A(c - d) = \{0\}$, then $\sigma_A(c(c - d)) \leq \sigma_A(c)\sigma_A(c - d) = 0$ by Corollary 5.7. Hence c(c - d), which is idempotent, has a negative which is q.r., and therefore c(c - d) = 0. Thus c = cd, and similarly d = cd; consequently c = d, which is impossible.

Q.E.D.

Theorem 10.3. _Let A be a commutative, complete, locally m-convex symmetric *-algebra with unit e, and suppose that $\mathcal{M}^-(A)$ is the union of two disjoint open subsets $\mathcal{M}_1(A)$ and $\mathcal{M}_0(A)$. Then there exists a unique idempotent element d in A such that f(d) = n if f $\in \mathcal{M}_n(A)$ (n = 0, 1)._

Proof. That at most one such d exists follows from Lemma 10.2 and Corollary 5.6 (b).

To show that at least one such d exists, let $\{U_i\}$ be an m-base for A. For each i, let $\mathcal{M}_n(\bar{A}_i)$ be the image in $\mathcal{M}^-(\bar{A}_i)$ of $\mathcal{M}_n(A) \cap U_i^o$, under the canonical homeomorphism $f \longrightarrow \bar{f}_i$ of Proposition 7.5 (n = 0, 1). Let us show that for every i, we can find a unique idempotent $d(i) \in \bar{A}_i$ such that $g(d(i)) = n$ whenever $g \in \mathcal{M}_n(\bar{A}_i)$ (n = 0, 1).

Consider a fixed i. That at most one such d(i) exists follows from Lemma 10.2. Let us show that at least one such d_i exists. If $\mathcal{M}_1(\bar{A}_i)$ is empty, we pick d(i) = 0; if $\mathcal{M}_0(\bar{A}_i)$ is empty, we pick $d(i) = e_i$. If neither $\mathcal{M}_0(\bar{A}_i)$ nor $\mathcal{M}_1(\bar{A}_i)$ is empty, then we proceed as follows: The proof of Satz 18 of [12] shows that the required d(i) exists, provided that B_i, the subalgebra of $C(\mathcal{M}^-(\bar{A}_i))$ generated by A_i, is dense in $C(\mathcal{M}^-(\bar{A}_i))$ in the "sup." norm. By [27, Corollary 2, p.238], we need only show that B_i is separating and closed under conjugation. That B_i is separating is clear; B_i is also closed under conjugation since, if $x_i \in A_i$ and $f \in \mathcal{M}(\bar{A}_i)$, then $f((x^*)_i) = f \circ \pi_i(x^*) = \overline{f \circ \pi_i(x)}$ (by Lemma 6.4 (b)) $= \overline{f(x_i)}$.

Using the notation developed before Theorem 5.1, let $j \geq i$; we will show that $\bar{\pi}_{ij}(d(j)) = d(i)$. By the uniqueness of d(i), we need only show that $\bar{\pi}_{ij}(d(j))$ is idempotent, and that $g[\bar{\pi}_{ij}(d(j))] = n$ whenever $g \in \mathcal{M}_n(\bar{A}_i)$ (n = 0, 1). Now since d(j) is idempotent, the same is true of $\bar{\pi}_{ij}(d(j))$. Also, if $g \in \mathcal{M}_n(\bar{A}_i)$, then $g[\bar{\pi}_{ij}(d(j))] = g \circ \bar{\pi}_{ij}(d(j)) = n$, since $g \circ \bar{\pi}_{ij} \in \mathcal{M}_n(\bar{A}_j)$ (n = 0, 1).

By Theorem 5.1 (b), we now have a $d \in A$ such that $d_i = d(i)$ for all i. Now d is certainly idempotent, since, for all i, $(d^2)_i = (d_i)^2 = (d(i))^2 = d(i) = d_i$. To complete the proof, we need only show that $f(d) = n$ for $f \in \mathcal{M}_n(A)$ $(n = 0, 1)$. But if $f \in \mathcal{M}_n(A)$, then, by Proposition 7.5 and the definitions, there exists an i and an $f_i \in \mathcal{M}_n(\bar{A}_i)$ such that $f(x) = f_i(x_i)$ for all $x \in A$; in particular, $f(d) = f_i(d(i)) = n$ $(n = 0, 1)$.

Q.E.D.

Corollary 10.4. <u>Let A be a commutative, complete, locally m-convex symmetric *-algebra with unit e. Then $\mathcal{M}^-(A)$ is disconnected if and only if A is the internal direct sum of two proper ideals with unit.</u>

Proof. If $\mathcal{M}^-(A)$ is disconnected, let d be the element of Theorem 10.4. Then, as is easily seen and shown in the proof of Satz 18 of [12], Ad and A(e - d) are ideals of the required sort. Suppose, conversely, that A is the internal direct sum of proper ideals with units c and d respectively. As in [12, Satz 17], we have $c^2 = c$, $d^2 = d$, and $c + d = e$; if $\mathcal{M}^-(A)$ were not disconnected, it would follow that $f(c) = 0$ for all $f \in \mathcal{M}(A)$ or $f(d) = 0$ for all $f \in \mathcal{M}(A)$, and we may suppose the former. But then $c = 0$ by Lemma 10.2 and Corollary 5.6(b), which is a contradiction.

Q.E.D.

11. Topological divisors of zero.

Definition 11.1. Let A be a topological algebra and let $x \in A$. Then $x + \lambda$ is a <u>strong topological divisor of 0</u> if either $R_{x+\lambda}$ or $L_{x+\lambda}$ is not an isomorphism into.

If A is a normed algebra, then this concept agrees with the concept of topological divisor of 0 due to Shilov. The generalization in Definition 11.1 is essentially due to Arens [3, p.627] who did <u>not</u> use the word <u>strong</u>. While the concept defined in Definition 11.1 is simple and aesthetically pleasing, we were unable to use it for the purpose of extending the relevant theorems for Banach algebras to locally m-convex algebras (Proposition 11.6 - Proposition 11.2). We therefore propose the following concept, which is less aesthetic, and which is meaningful only for locally m-convex algebras, but for which the above theorems do go through. Proposition 11.3 tells what the author knows about the relationship between the two concepts; the author does not know whether the two concepts are actually equivalent.

Definition 11.2. Let A be locally m-convex algebra, and let $x \in A$. Then $x + \lambda$ is a *topological divisor of 0* in A if, whenever $\{U_i\}$ is an m-base for A, there exists an i such that $x_i + \lambda$ is a strong topological divisor of 0 in $\overline{A}_i$.

Remark. Observe that every algebraic divisor of 0 is also a topological one.

Remark. Observe that the following are equivalent:

a) $x_i + \lambda$ is a strong topological divisor of 0 in $\overline{A}_i$.
b) $x_i + \lambda$ is a strong topological divisor of 0 in A_i.
c) $(R_{x+\lambda})_i$ or $(L_{x+\lambda})_i$ is not an isomorphism into (see Definition (A.1).

Proposition 11.3. *Let A be a (normed) locally m-convex algebra, and let* $x \in A$. *Then* $x + \lambda$ *is a topological divisor of* 0 *if (and only if)* $x + \lambda$ *is a strong topological divisor of* 0.

Proof. Suppose that $x + \lambda$ is not a topological divisor of 0. Then we can find an m-base $\{U_i\}$ for A such that, for every i, $(R_{x+\lambda})_i$ and $(L_{x+\lambda})_i$ are isomorphisms into (see Def. A.1). It follows, by Theorem A.1 (c), that $R_{x+\lambda}$ and $L_{x+\lambda}$ are isomorphisms into, whence $x + \lambda$ is not a strong topological divisor of 0.

The converse for *normed* algebras A (see comment after Proposition 2.3) is clear, since if S is the unit sphere of A, then $\{S\}$ is clearly an m-base for A.

Q.E.D.

Remark. It follows from Proposition 11.3 that the concepts of "topological divisor of 0" and "strong topological divisor of 0" agree for *normed* algebras. We will henceforth omit the word "strong" when dealing with normed algebras.

Proposition 11.4. *If A is a locally* m-*convex algebra, if* $x \in A$, *and if* x *(resp.* x + 1) *is a topological divisor of* 0, *then* x *is not invertible (resp. q.r.).*

Proof. Let $\{U_i\}$ be an m-base for A. Then, for some i, x_i (resp. $x_i + 1$) is a topological divisor of 0 in $\overline{A}_i$. Hence, for this i, x_i is not invertible (resp. q.r.) in $\overline{A}_i$, and therefore x cannot be invertible (resp. q.r.).

Q.E.D.

Proposition 11.5. *Let $\hat{A}$ be a locally m-convex algebra, A a (dense) subalgebra of $\hat{A}$, and $x \in A$. Then $x + \lambda$ is a topological divisor of 0 in $\hat{A}$ if (and only if) $x + \lambda$ is one in A.*

Proof. Following the notation developed before Proposition 7.8, we will equip everything connected with $\hat{A}$ with a hat.

Suppose that $x + \lambda$ is a topological divisor of 0 in A, and let $\{\hat{U}_i\}$ be an m-base for $\hat{A}$. If $U_i = \hat{U}_i \cap A$, then, by definition, $\{U_i\}$ is an m-base for A. Hence there exists an i such that $\pi_i(x) + \lambda$ is a topological divisor of 0 in A_i. It follows from Proposition 7.8 that $\hat{\pi}_i(x) + \lambda$ is a topological divisor of 0 in $\hat{\pi}_i(A) \subset \hat{A}_i$. Hence $\hat{\pi}_i(x) + \lambda$ is a topological divisor of 0 in $\hat{A}_i$, and therefore $x + \lambda$ is a topological divisor of 0 in $\hat{A}$.

Suppose now that A is dense in $\hat{A}$, and that $x + \lambda$ is a topological divisor of 0 in $\hat{A}$. Let $\{U_i\}$ be an m-base for A. If $\hat{U}_i$ is the closure of U_i in $\hat{A}$, then $\{\hat{U}_i\}$ is an m-base for $\hat{A}$.(6) Hence for some i, $\hat{\pi}_i(x) + \lambda$ is a topological divisor of 0 in $\hat{A}_i$, and therefore [18, Lemma 3.4] in the dense subalgebra $\hat{\pi}_i(A)$. Again using Lemma 7,8, it follows that $\hat{\pi}_i(x) + \lambda$ is a topological divisor of 0 in A_i, and hence $x + \lambda$ is a topological divisor of 0 in A.

Q.E.D.

Propositions 11.3 and 11.4 show that the properties which *strong* topological divisors of 0 have are shared by our topological divisors of 0. The following propositions will show that conditions which in a Banach algebra are sufficient for an element to be a topological divisor of 0 are also sufficient in complete locally m-convex algebras.

Proposition 11.6. *Let A be a complete, locally m-convex algebra, and let $x \in A$. If x is in the closure of the set of q.r. elements of A but is not itself q.r., then $1 + x$ is a topological divisor of 0.*

Proof. Let $\{U_i\}$ be an m-base for A. Clearly x_i is in the closure of the q.r. elements of A_i, and a fortiori those of $\bar{A}_i$, for all i. By Theorem 5.2, on the other hand, x_{i_o} is not q.r. in A_{i_o} for some i_o. Hence, by

(6) This follows from the following fact: If X is a topological space, Y a dense subspace of X, $y \in Y$, and U a neighborhood of y in Y, then the closure of U in X is a neighborhood of y in X.

[18, Lemma 3.3], $1 + x_{i_0}$ is a (two sided !) topological divisor of 0 in $\bar{A}_{i_0}$. It follows that $1 + x$ is a topological divisor of 0 in A.

Q.E.D.

Remark. The proof of Proposition 11.6 shows that, if $\{U_i\}$ is an m-base for A, then there exists an i such that $1 + x_i$ is a two-sided topological divisor of 0 in $\bar{A}_i$. This assertion carries over to similar assertions about some of the later propositions.

Proposition 11.7. Let A be a complete, locally m-convex algebra, and let $x \in A$. If $-\lambda x$ is q.r. for arbitarily large λ, then either x is a topological divisor of 0, or x is invertible.

Proof. Suppose that x is not a topological divisor of 0. Then there exists an m-base $\{U_i\}$ for A such that x_i is not a topological divisor of 0 for any i. Also, for every i, $-\lambda x_i$ is q.r. in $\bar{A}_i$ for arbitrarily large λ. Hence, by [24, Theorem 1.5], (7) x_i is invertible in $\bar{A}_i$ for all i. Consequently, by Theorem 5.2 (c), x is invertible in A, as we had to show.

Q.E.D.

Proposition 11.8. Let A be a complete, locally m-convex algebra, and let $x \in A$. Let $\lambda \in \Sigma_A(x)$ be a boundary point of $\Sigma_A(x)$. Then $x - \lambda$ is a topological divisor of 0.

Proof. $\lambda \neq 0$: In this case, $-\lambda^{-1}x$ satisfies the conditions of Proposition 11.6. Hence $-\lambda^{-1}x + 1$ is a topological divisor of 0, and therefore so is $x - \lambda$.

$\lambda = 0$: In this case, x satisfies the conditions of Proposition 11.7. Since $0 \in \Sigma_A(x)$, x is, by definition, not invertible. It follows from Proposition 11.7 that x is a topological divisor of 0.

Q.E.D.

From Proposition 11.8, Proposition 11.5, and Proposition 11.3 we conclude the following

Proposition 11.9. Let $\hat{A}$ be a locally m-convex algebra with complete subalgebra A, and let $x \in A$. If $\lambda \in \Sigma_A(x)$, and if λ is a boundary point

---

(7) Rickart talks about λx instead of $-\lambda x$ in [24] since for him $x \circ y = x + y - xy$.

of $\Sigma_A(x)$, *then* $\lambda \in \Sigma_{\hat{A}}(x)$.

Proof. By Proposition 11.8, $x - \lambda$ is a topological divisor of 0 in A, and hence $x - \lambda$ is one in $\hat{A}$ by Proposition 11.5. If $\lambda \neq 0$, it follows that $- {}^{-1}x + 1$ is a topological divisor of 0 in $\hat{A}$; hence (Proposition 11.4) $-\lambda^{-1}x$ is not q.r. in $\hat{A}$, and therefore $\lambda \in \Sigma_{\hat{A}}(x)$. If $\lambda = 0$, then (Proposition 11.4) x is not invertible in $\hat{A}$, and hence again $\lambda \in \Sigma_{\hat{A}}(x)$.

Q.E.D.

Corollary 11.10. *Let A be a symmetric *-algebra which is a complete subalgebra of a locally m-convex algebra* $\hat{A}$, *and let* $x \in A$. *If x is q.r. in* $\hat{A}$, *then x is q.r. in A.*

Proof. Let $y = x \circ x^*$; then y is q.r. in $\hat{A}$, and therefore $-1 \notin \Sigma_{\hat{A}}(y)$. Also y is self adjoint, and therefore (Proposition 6.4 (a)) $\Sigma_A(y)$ is real. It follows, by Proposition 11.9, that $\Sigma_A(y) = \Sigma_{\hat{A}}(y)$. Hence $-1 \notin \Sigma_A(y)$, and therefore y is q.r. in A. Similarly $x^* \circ x$ is q.r. in A, and hence x is q.r. in A.

Q.E.D.

So far we did not need the spectra of the elements of A to be compact. In the remainder of this section we will need it, and we will therefore deal only with Q-algebras (Appendix E).

Proposition 11.11. *Let A be a locally m-convex algebra whose completion* $\bar{A}$ *is a Q-algebra, and let* $x \in A$. *Then* $x - \lambda$ *is a topological divisor of 0 for some* λ.

Proof. By Proposition 11.5, we may assume that $A = \bar{A}$. Since $\Sigma_A(x)$ is compact, there exists a $\lambda \in \Sigma_A(x)$ which is a boundary point of $\Sigma_A(x)$. Then this λ does the trick, by Proposition 11.8.

Q.E.D.

Using arguments of [18, Theorem 3.1] we now prove

Proposition 11.12. *Let A be a locally, m-convex algebra containing no non-zero topological divisors of 0, and whose completion* $\bar{A}$ *is a Q-algebra. Then A is isomorphic to the algebra of complex numbers.*

Proof. Let $x \in A$, $x \neq 0$. By Proposition 11.11, there exists a scalar λ such that $x - \lambda$ is a topological divisor of 0 in $\bar{A}$. By Proposition 11.4, x cannot be a topological divisor of 0 in $\bar{A}$, and therefore $\lambda \neq 0$. Letting

$y = -\lambda^{-1}x$, it follows that $y + 1$ is a topologiviser of 0 in $\bar{A}$, and hence in A (Proposition 11.4). If $z \varepsilon A$, then it is easily seen that $w = [(y + 1)z][z(y + 1)]$ is a topological divisor of 0 in A; since $w \varepsilon A$, it follows that $w = 0$. Hence one of the factors of w must also be 0; suppose $(y + 1)z = 0$. Then $[z(y + 1)]z = 0$; if $z \neq 0$, it follows that $z(y + 1)$ is a topological divisor of 0 and hence is 0, and if $z = 0$, then $z(y + 1)$ is certainly 0. Thus $(y + 1)z = z(y + 1) = 0$, and hence $-y$ is a unit for A. Thus every $x \varepsilon A$ is a scalar multiple of the unit, and hence A is isomorphic to the algebra of complex numbers.

Q.E.D.

12. Continuity of multiplicative linear functionals. In a Banach algebra, or more generally in a Q-algebra (Appendix E), all regular maximal ideals are closed. This need no longer be true in arbitrary complete locally m-convex algebras; in fact, it is false in the commutative $\mathfrak{F}$-algebra of Example 3.1. However, it can be shown (see, for instance, [14, Theorem 50 (modified for complex case)] or Proposition 13.5) that in this algebra all regular maximal ideals, modulo which we get the algebra of complex numbers, are closed. In other words, every multiplicative linear functional on this algebra is continuous; the purpose of this section is to determine what algebras have this property. To facilitate our discussion, we make the following

Definition 12.1. A topological algebra A is called functionally continuous if every multiplicative linear functional on A is continuous.

Throughout this section, we will restrict our attention to commutative algebras. In general, only complete algebras can be expected to be functionally continuous; this is already evident in the normed case, where the algebra of polynomials, with $||x|| = \sup_{0 \leq t \leq 1} |x(t)|$, is not functionally continuous. Once we have a complete, commutative locally m-convex algebra, there is already a considerable connection between the set of all multiplicative linear functionals and the set of continuous ones; the extent of our knowledge in this case is summed up in Lemma 6.1 (b). Such algebras need not, however, be functionally continuous, as is shown by parts (a) and (b) of the following

Proposition 12.2. *Let T be a countably compact, non-compact, completely regular space, which is either locally compact or first countable. Let A = C(T) with the compact-open topology. Then*

a) *A is a complete, commutative, locally m-convex algebra.*

b) *A is not functionally continuous.*

c) *Every multiplicative linear functional on A is bounded (i.e., sends bounded sets into bounded sets).*

Proof.

a) This follows from Example D.2, Example D.3, and Lemma D.5.

b) By Example 7.6, every continuous multiplicative linear functional is generated by a point of T. Since T is countably compact, every x in C(T) is bounded. Hence, as is easy to see, the set of multiplicative linear functionals is precisely the set of functionals generated by the points of $\beta(T)$, the Stone-Čech compactification of T. Since T is not compact, (b) follows.

c) Let B be a bounded subset of A. If we let $x(t) = \sup_{y \in B} |y(t)|$, then it follows that x is bounded on every compact subset of T. In view of Lemma 6.1 (a), we will have proved (c) if we can show that x is bounded on T. Suppose x were not bounded on T; we can then find a sequence $\{t_n\}_{n=1}^{\infty}$ of distinct points of T such that $f(t_n) > n$ $(n = 1, 2, \ldots)$. Since T is countably compact, $\{t_n\}$ has a limit point t. Our assumptions now imply that we can find an infinite subset of $\{t_n\}$ whose closure in T is compact. Since x is not bounded on this set, we have a contradiction.

Q.E.D.

Remark. As a simple example of a space T satisfying the assumptions of Proposition 12.2, let T be the set of ordinals smaller than Ω, the first uncountable ordinal, in the order topology (see Example 3.7). It is known that $\beta(T)$ consists of T with Ω itself added. The functional generated by Ω sends every $x \in C(T)$ into that constant value which it must have "from some point on".

Observe that a space T satisfying the conditions of Proposition 12.2 is certainly not σ-compact, and hence, by Example 7.6 and Lemma 6.2 (c), A is not metrizable. This observation suggests the first of the following questions, while Proposition 12.2(c) suggests the second.

Question 1. Is every commutative $\mathcal{F}$-algebra functionally continuous?

Question 2. Is every multiplicative linear functional on a commutative, complete, locally m-convex algebra bounded?

We cannot answer either of these questions completely. (Observe that an affirmative answer to Question 2 implies an affirmative answer to Question 1, since every bounded linear functional on a metrizable, locally convex space is continuous.) We can, however, give a fairly satisfactory partial answer to Question 1 (in Theorem 12.6 and Theorem 12.7 below), and the remainder of this section will be devoted to obtaining the answer.

We begin with the following

Lemma 12.3. *Let A be a locally m-convex algebra.*

a) *If A is commutative and complete (which implies (Corollary 5.5) that $\mathcal{R}(A)$ is closed), and if $A/\mathcal{R}(A)$ is functionally continuous, then so is A.*

b) *If A is an ideal in a locally m-convex algebra B, and if B is functionally continuous, then so is A.*

c) *If A^+ is functionally continuous, then so is A.*

d) *If A is the convex hull (see Definition 15.1) of functionally continuous algebras, then A is functionally continuous.*

Proof.

a) Let f be a multiplicative linear functional on A. Then f vanishes on $\mathcal{R}(A)$, and hence there exists a multiplicative linear functional g on $A/\mathcal{R}(A)$ such that $f = g \circ \pi$, where π is the projection from A to $A/\mathcal{R}(A)$. But g must be continuous, and hence so is f.

b) Let f be a multiplicative linear functional on A. By Theorem C.1, f can be extended to a multiplicative linear functional g on B. But g must be continuous, and hence so is f.

c) This follows from (b).

d) Let f be a multiplicative linear functional on A. Then the restriction of f to each generating algebra is continuous by assumption, so f is continuous.

Q.E.D.

Lemma 12.3 reduces our problem for commutative and complete locally m-convex algebras to the case where they are semi-simple and have a unit. By Proposition 8.1, such an algebra A is isomorphic to an algebra of continuous complex valued functions, containing the identity function, on the completely regular space $\mathcal{M}(A)$; if the algebra is metrizable, then this space is σ-compact (Lemma 6.2(c)) and hence Lindelöf. We now turn to such function algebras.

Proposition 12.5. Let T be a completely regular space, and let A be an algebra of continuous, complex-valued functions, containing the identity function e, on T, such that the inverse function of every element of A which vanishes nowhere is also in A. Then every non-zero multiplicative linear functional on A is generated by a point of T, provided A satisfies conditions (I) and either (II) or (II') below.

I. If $x_1, \ldots, x_n$ are elements of A such that $Z(x_1) \cap \ldots \cap Z(x_n) \neq \emptyset$,(8) then there are elements $y_1, \ldots, y_n$ in A such that $x_1y_1 + \ldots + x_ny_n = e$.

II. There exists a finite set $\{a_1, \ldots, a_m\}$ of elements of A such that $\bigcap_{n=1}^{m} \{t \in T | a_n(t) = \lambda_n\}$ is compact for any m scalars $\lambda_1, \ldots, \lambda_m$.

II'.

a) T is a Lindelöf space.

b) If $\{x_n\}_{n=1}^{\infty}$ is a sequence of elements of A, then there exists a sequence $\{y_n\}_{n=1}^{\infty}$ of elements of A such that

b_1) $Z(x_n) = Z(y_n)$ $(n = 1, 2, \ldots)$

b_2) $|y_n(t)| \leq 1$ for all $t \in T$ $(n = 1, 2, \ldots)$

b_3) If $y(t) = \sum_{n=1}^{\infty} \frac{1}{2^n} y_n(t)$, then $y \in A$.

---

(8) Following Hewitt [14], we let $Z(x) = \{t \in T | x(t) = 0\}$. We also use $\emptyset$ to denote the null set.

Proof. Let f be a non-zero multiplicative linear functional, and let M be its kernel. We will frequently use the fact that, since M is an ideal $\neq$ A, and since every non-vanishing element of A has an inverse in A, M cannot contain any non-vanishing elements. We now begin our proof by establishing four consequences of (I):

(1) If $x_1, \ldots, x_n$ are elements of M, then $Z(x_1) \cap \ldots \cap Z(x_n) \neq \emptyset$: For if not, it would follow from (I) that M has a non-vanishing element.

(2) For every $x \in A$, $f(x) = x(t)$ for some $t \in T$: If not, then $x - f(x)e$ would be a non-vanishing element of M.

(3) If τ is the topology on A of uniform convergence on T, then f is continuous with respect to τ: If $|x(t)| < \varepsilon$ for all $t \in T$, then $|f(x)| < \varepsilon$ by (2).

(4) If $x \in M$, $y \in A$, and $Z(x) \subset Z(y)$, then $y \in M$: If not, then x and $y-f(y)e$ are elements of M such that $Z(x) \cap Z(y-f(y)e) = \emptyset$, contradicting (1).

Suppose now that f is not generated by an element of T. Then clearly

$$(*) \qquad \bigcap_{x \in M} Z(x) = \emptyset .$$

We will show that, assuming either II or II', this leads to a contradiction.

II: If $a_1, \ldots, a_m$ are as in (II), then the functions $a_i - f(a_i)e$ are all in M, and $\bigcap_{i=1}^{m} Z(a_i - f(a_i)e)$ is a compact set which we will call C. Now by (1), the collection $\{C \cap Z(x)\}_{x \in M}$ of closed subsets of C has the finite intersection property, and hence has a non-empty intersection. But by (*), $\bigcap_{x \in M} C \cap Z(x) = \bigcap_{x \in M} Z(x) = \emptyset$, which is a contradiction.

II': By (*) and II(a), there exists a sequence $\{x_n\}_{n=1}^{\infty}$ of elements of M such that $\bigcap_{n=1}^{\infty} Z(x_n) = \emptyset$. Now let $\{y_n\}_{n=1}^{\infty}$ and y be as in II'(b). Then clearly y never vanishes. On the other hand, $f(y_i) = 0$ for all i (by (4)), and since $\sum_{n=1}^{N} y_n(t) \xrightarrow{N} y(t)$ uniformly on T, it follows (by (3)), that $f(y) = 0$, which is impossible.

Q.E.D.

Remark. If the conjugate function of every $x \in A$ is also in A, then (I) is certainly satisfied, since we may simply let $y_i = \bar{x}_i / (x_1\bar{x}_1 + \ldots + x_n\bar{x}_n)$.

We are now ready to state and prove the principal theorems of this section.

Theorem 12.6. Let A be a commutative $\mathcal{F}$-algebra which is also a symmetric *-algebra. Then A is functionally continuous.

Proof. By Lemma 12.3 and Lemma 6.7, we need only prove the theorem for the case where A is semi-simple and has a unit. Proposition 8.1, Lemma 6.2(c), and Theorem 5.4 show that in this case, A is algebraically isomorphic to an algebra of functions, containing the identity function, on the σ-compact, completely regular space $\mathcal{M}^-(A)$, such that the inverse of every function which vanishes nowhere is also in the algebra. Let us identify A with this function algebra. We are therefore in the situation of Proposition 12.5, and to complete the proof of the theorem, we need only verify conditions (I) and (II') of Proposition 12.5.

By Corollary 6.5, for every x in A we have $x^*(t) = \overline{x(t)}$ for all $t \in \mathcal{M}^-(A)$. Hence, by the remark following Proposition 12.5, (I) is satisfied. Since every σ-compact space is Lindelöf, II'(a) is also satisfied. It remains to verify II'(b). Let $\{x_n\}_{n=1}^{\infty}$ be a sequence of elements of A. By [10, Proposition 3], we can find scalars λ_n (n = 1, 2, ...), $0 \leq \lambda_n \leq 1$, such that $\{\lambda_n(x_nx_n^*)(1 + x_nx_n^*)^{-1}\}_{n=1}^{\infty}$ is a bounded subset of A. Let $y_n = \lambda_n(x_nx_n^*)(1 + x_nx_n^*)^{-1}$. Then, for all $t \in \mathcal{M}^-(A)$, $y_n(t) = \lambda_n|x_n(t)|^2(1 + |x_n(t)|^2)^{-1}$, and hence $\{y_n\}_{n=1}^{\infty}$ certainly satisfies II'(b_1) and II'(b_2). Let us show that II'(b_3) also holds. By Lemma 7.7(b), the series $\sum_{n=1}^{\infty} \frac{1}{2^n} y_n$ converges to some $y \in A$. Since $x \longrightarrow x(t)$ is a continuous functional on A for each $t \in \mathcal{M}^-(A)$, it follows that $y(t) = \sum_{n=1}^{\infty} \frac{1}{2^n} y_n(t)$ for all $t \in \mathcal{M}^-(A)$, which proves II'(b_3).

Q.E.D.

In the next theorem we will consider a class of $\mathcal{F}$-algebras which are far from being symmetric.

Theorem 12.7. Let T be connected, open subset of the complex plane, or, more generally, a connected, open subset of the space of n complex variables whose associated domain of regularity [8, p.624] is univalent. Let A be the algebra of functions which are analytic on T, in the compact open topology. Then A is a functionally continuous $\mathcal{F}$-algebra.

Proof. That A is an $\mathcal{F}$-algebra is easy to see, and was already asserted in Example 3.3. Let us first suppose that T is a connected, open subset of the complex plane, or, more generally, a connected, open subset of the space of n-complex variables which is a domain of regularity. Since the functionals generated by the points of T are obviously continuous, it is sufficient to show that every multiplicative linear functional on A is generated by a point of T. This puts us into the situation of Proposition 12.5, and we need only check that conditions (I) and (II) of that proposition hold:

I. If the dimension of the space is m, let $a_k(\zeta_1, \ldots, \zeta_m) = \zeta_k$ $(k = 1, \ldots, m)$.

II. This is precisely the assertion of [7, pg.63, Corollaire]. (For the case where T is the complex plane, this result already follows from [29, Theorem 9]).

Now if T is arbitrary, let $\tilde{T}$ be the domain of regularity associated with T, and let $\tilde{A}$ be the algebra of functions which are analytic on $\tilde{T}$ in the compact-open topology. Let θ be the natural mapping from $\tilde{A}$ to A. By assumption, θ is an algebraic isomorphism onto, and θ is certainly continuous. Since A and $\tilde{A}$ are both $\mathcal{F}$-spaces, it follows from the closed graph theorem [4, Th 5, pg.41] that θ^{-1} is also continuous. Hence A and $\tilde{A}$ are isomorphic, and the problem has been reduced to the previous case.

Q.E.D.

13. Elements with bounded spectrum. The main purpose of this section is to show (Theorem 13.6 and Theorem 13.7 below) that, under certain conditions, an algebra all of whose elements have bounded spectrum is a Q-algebra (Appendix E).

Notation 13.1. If A is an algebra, then

$$\mathcal{B}(A) = \{x \in A \mid \sigma_A(x) < \infty\},$$

$$\mathcal{S}(A) = \{x \in A \mid \sigma_A(x) \leq 1\}.$$

Remark. $A = \mathcal{B}(A)$ if and only if every $x \in A$ is a scalar multiple of some element of $\mathcal{S}(A)$.

Lemma 13.2. $\mathcal{B}(A)$ is closed under multiplication by scalars; if A is a commutative and complete locally m-convex algebra, then $\mathcal{B}(A)$ is a subalgebra of A, and $\mathcal{S}(A)$ is m-convex.

Proof. The first assertion is obvious, and the second follows from Corollary 5.7.

Q.E.D.

Let us see what the algebra $\mathcal{B}(A)$ can look like in some particular commutative and complete locally m-convex algebras.

Example 13.3.

a) In Example 3.4, Example 3.5, and 3.6, $\mathcal{B}(A) = A$.

b) In Example 3.1, $\mathcal{B}(A)$ is dense in A, but $\mathcal{B}(A) \neq A$.

c) In Example 3.2, $\mathcal{B}(A)$ contains only the constant functions.

We now proceed to establish some connections with Q-algebras.

Lemma 13.4. Let A be an algebra, and U a convex and symmetric subset of A which contains only q.r. elements. Then $U \subset \mathcal{S}(A)$.

Proof. Suppose not. Then there is an $x \in U$ such that $-\lambda^{-1}x$ is not q.r. for some λ with $|\lambda| < 1$. Since $|\lambda| < 1$, $-\lambda^{-1}x$ is also in U, which is impossible.

Q.E.D.

Proposition 13.5. Let A be a locally convex topological algebra. Then the following assertions are equivalent.

a) A is a Q-algebra.

b) $\mathcal{S}(A)$ is a neighborhood of 0 (i.e., 0 is in the interior of $\mathcal{S}(A)$).

c) $\mathcal{S}(A)$ has an interior.

Proof.

a $\longrightarrow$ b: This follows from Lemma 13.4.

b $\longrightarrow$ c: Obvious.

c $\longrightarrow$ a: If $\mathcal{S}(A)$ has an interior, so does $\frac{1}{2}\mathcal{S}(A)$. But every element of $\frac{1}{2}\mathcal{S}(A)$ is q.r., so the set of q.r. elements has an interior. Hence A is a Q-algebra by Lemma E.2.

Q.E.D.

Theorem 13.6. *Let A be a commutative, complete, locally m-convex algebra which has property (g) of Proposition 4.2. (in particular, A may be a commutative $\mathcal{F}$-algebra). Then the following assertions are equivalent.*

a) *A is a Q-algebra.*

b) $A = \mathcal{B}(A)$.

c) $\mathcal{M}(A)$ *is compact.*

Proof. Observe first that, by Corollary 5.6 (b), we have

$$(*) \qquad \sigma_A(x) = \sup\{|f(x)| \;|\; f \in \mathcal{M}(A)\}$$

and hence

$$(**) \qquad \mathcal{S}(A) = \bigcap_{f \in \mathcal{M}(A)} \{x \in A \,|\, |f(x)| \leq 1\} = [\mathcal{M}(A)]^{o} .$$

a $\rightarrow$ b: This follows from Lemma E.3.

a $\rightarrow$ c: By Proposition 13.5, $\mathcal{S}(A)$ is a neighborhood of 0, and hence, as is well known, $[\mathcal{S}(A)]^{o}$ is weakly compact. By (**), $\mathcal{M}(A) \subset [\mathcal{M}(A)]^{oo} = [\mathcal{S}(A)]^{o}$; since $\mathcal{M}(A)$ is weakly closed in A' (Lemma 6.2 (a)), $\mathcal{M}(A)$ is weakly compact.

c $\rightarrow$ b: This follows from (*).

b $\rightarrow$ a: By (**), $\mathcal{S}(A)$ is closed, symmetric and m-convex, and (a) implies that $\mathcal{S}(A)$ spans A. By assumption it follows that $\mathcal{S}(A)$ is a neighborhood of 0, and hence A is a Q-algebra by Proposition 13.5.

Q.E.D.

Remark. Observe that the assumption that A has property (g) of Proposition 4.2 cannot be dropped. Consider, in fact, the algebras of Example 3.7 and Example 3.8, both of which are commutative, complete, and locally m-convex; the algebra of Example 3.8 satisfies assertions (b) and (c) of Theorem 13.6, but not assertion (a); the algebra of Example 3.7 satisfies assertion (b), but not assertions (a) or (c).

In the proof of Theorem 13.6, b $\rightarrow$ a, we made essential use of the fact that $\mathcal{S}(A)$ was convex and closed. If A is not commutative, then $\mathcal{S}(A)$ need not be convex, but we nevertheless have

Theorem 13.7. *Let A be an $\mathcal{F}$-algebra such that $\mathcal{S}(A)$ is closed. Then the following assertions are equivalent.*

a) *A is a Q-algebra.*

b) $A = \mathfrak{R}(A)$.

Proof.

a $\longrightarrow$ b: Just as in Theorem 13.6, this follows from Lemma E.3.

b $\longrightarrow$ a: By the remark following Notation 13.1, $A = \bigcup_{n=1}^{\infty} n\,\mathcal{S}(A)$. Since $\mathcal{S}(A)$ is closed, and A is of the second category, $n\,\mathcal{S}(A)$ must have an interior for some n. Hence $\mathcal{S}(A)$ has an interior, and hence A is a Q-algebra by Proposition 13.5.

Q.E.D.

In Theorem 13.7, the assumption that $\mathcal{S}(A)$ is closed played an important role. We do not know whether this assumption can be omitted, so we now conclude this section with a lemma which gives sufficient conditions for $\mathcal{S}(A)$ to be closed.

Lemma 13.8. *If A is a locally m-convex algebra, then each of the following conditions is sufficient for A to be closed.*

a) *A is commutative.*

b) *For some m-base* $\{U_i\}$ *for A, every* $\mathcal{S}(\overline{A}_i)$ *is closed in* $\overline{A}_i$.

c) *A has property (a) of Proposition 9.2 and, for every primitive ideal P associated with some closed regular maximal one sided ideal in A,* $\mathcal{S}(A/P)$ *is closed in A/P.*

d) *A satisfies a polynomial identity or is completely continuous.*

Proof.

a) This follows from (**) of the proof of Theorem 13.6, as we already observed in that proof.

b) Since (Corollary 5.3(a)) $\Sigma_A(x) = \bigcup_i \Sigma_{\overline{A}_i}(x_i)$, it follows that $\mathcal{S}_A = \bigcap_i \pi_i^{-1}(\mathcal{S}_{\overline{A}_i})$.

c) Let $\mathcal{P}$ be the collection of all primitive ideals in A which are associated with some closed, regular, maximal one sided ideal in A. Property (a) of Proposition 9.2 implies that, for $x \in A$, x is q.r. if and only if x is q.r. mod. every $P \in \mathcal{P}$. Hence $\Sigma_A(x) = \bigcup_{P \in \mathcal{P}} \Sigma_{A/P}(x_p)$, where x_p denotes the image of x in A/P. Hence $\mathcal{S}(A)$ is closed, just as for (b).

d) In this case, it follows from Theorem 9.8, Proposition 9.6(a), and Theorem 9.3 that A satisfies condition (a) of Proposition 9.2. Let $\mathcal{P}$ be the collection of the proof of (c). If $P \in \mathcal{P}$, then, by Proposition B.8 and Lemma B.14, A/P is finite dimensional, and hence $\mathcal{S}(A/P)$ is finite and therefore closed.

Q.E.D.

Remark. Condition (b) (and sometimes condition (c)) reduces the problem of whether $\mathcal{S}(A)$ is closed to the case where A is a Banach algebra. In this case, $\mathcal{S}(A)$ is certainly closed if σ_A is continuous, which is in turn the case if the mapping $x \longrightarrow \Sigma_A(x)$, with the Hausdorff metric on the space of compact subsets of the complex plane, is continuous. This latter question is discussed in [23].

Remark. Kakutani has shown (unpublished) that $\mathcal{S}(A)$ is not closed if A is the Banach algebrá of all bounded operators or an infinite dimensional Hilbert space.

14. <u>Uniqueness of topology.</u> One of the prettiest theorems in the theory of Banach algebras [12, Satz 17] can be phrased as follows: If A is a semi-simple, commutative algebra then A can be given at most one topology which makes it into a Banach algebra. (This theorem has been generalized by Rickart [24]; it is still an unsolved problem whether the conclusion of the theorem is true for <u>all</u> semi-simple algebras.) The purpose of the section is to extend this theorem (and one of its generalizations in [24]) from Banach algebras to more general $\mathcal{F}$-algebras (since the closed-graph theorem plays an essential role in the proof, $\mathcal{F}$-algebras appear to be the approriate kind of extension).

In our approach, we will follow Rickart [24]. We begin by observing that Lemma 2.1, Lemma 2.2, and Theorem 2.3 of [24] remain true, with identical proofs, for $\mathcal{F}$-spaces instead of Banach spaces; the norm is now an $\mathcal{F}$-space norm, which satisfies the triangle inequality, but is not homogeneous with respect to scalars. As an immediate corollary to the $\mathcal{F}$-space version of Theorem 2.3 of [24], we have the following sharpened version of Theorem 2.10 of [24]:

Lemma 14.1. <u>Let A be a semi-simple algebra, and let τ_1 and τ_2 be two topologies which make A into an $\mathcal{F}$-algebra. Let $\mathcal{P}$ be a collection of primitive ideals in A, each of which is closed with respect to both τ_1 and τ_2, and whose intersection is 0. For each P ε $\mathcal{P}$, let $\tau_n(P)$ be the topology induced by τ_n on the factor algebra A/P (n = 1, 2). Then $\tau_1 = \tau_2$, provided $\tau_1(P) = \tau_2(P)$ for every P ε $\mathcal{P}$.</u>

We are now ready for our main theorems.

Theorem 14.2. <u>Let A be a commutative, semi-simple, functionally continuous (Definition 12.1) $\mathcal{F}$-algebra, with topology τ. Then τ is the only topology on A which makes A into an $\mathcal{F}$-algebra.</u>

Proof. Let τ_1 be any topology on A which makes A into an $\mathcal{F}$-algebra, and let $\mathcal{P}$ be the collection of kernels of multiplicative linear functionals on A which are τ_1-continuous. Since $\mathcal{R}(A) = 0$, it follows from Corollary 5.5 that the intersection of these kernels is 0. Now every P in $\mathcal{P}$ is τ_1-closed by definition of $\mathcal{P}$; it is also τ-closed, since A, with the topology τ, is functionally continuous. Our conclusion now follows from Lemma 14.1, when we observe that A/P is isomorphic to the complexes for every P ε $\mathcal{P}$.

Q.E.D.

Theorem 14.3. <u>Let A be a semi-simple algebra, satisfying either of the following two conditions:</u>

a) <u>A is commutative and symmetric.</u>

b) <u>A satisfies a polynomial identity, and every element of A has a bounded spectrum.</u>

<u>Then there is at most one topology which makes A into an $\mathcal{F}$-algebra.</u>

Proof.

a) This follows immediately from Theorem 14.2, since (Theorem 12.6) A is functionally continuous under any topology which makes it into an $\mathcal{F}$-algebra.

b) Let τ_1 and τ_2 be two topologies which make A into an $\mathcal{F}$-algebra. Observe, to begin with, that τ_1 and τ_2 make A into a Q-algebra (by Theorem 13.7 and Lemma 13.8(d)). Hence, if $\mathcal{P}$ is the collection of <u>all</u> primitive ideals in A, then every P ε $\mathcal{P}$ is both τ_1-closed and τ_2-closed. Furthermore, the intersection of all

$P \in \mathcal{P}$ is certainly 0. Our conclusion now follows from Lemma 14.1, if we observe (Lemma B.17) that A/P is finite dimensional for every $P \in \mathcal{P}$, and that every n-dimensional topological linear space is isomorphic to Euclidean n-space.

Q.E.D.

Remark. If the answer to Question 1 of Section 12 is affirmative, then, as the proof shows, we could drop the assumption of symmetry in (a). This would, of course, be desirable, particularly since no such assumption is required in the case of Banach algebras.

As far as the author knows, the following corollary has never been explicitly stated even in the case where "$\mathcal{F}$-algebra" is replaced by "Banach algebra".

Corollary 14.4. Let A *be a semisimple Banach algebra, with topology* τ, *which satisfies a polynomial identity. Then* τ *is the only topology on* A *which makes* A *into an* $\mathcal{F}$*-algebra*.

Proof. Every element of a Banach algebra has a bounded spectrum.

Q.E.D.

It seems likely that some of Rickart's generalizations in [24] can be extended to $\mathcal{F}$-algebras, but that would take us beyond the scope of this paper.

15. $\mathcal{LF}$-algebras.

Definition.15.1. A sequence $\{E_n\}$ of locally convex spaces, with topologies τ_n, is called *compatible* if, for every n, we have

a) E_n is complete.

b) E_n is a proper subspace of E_{n+1}.

c) $\tau_{n+1}|E_n = \tau_n$.

The *convex hull* of a compatible sequence $\{E_n\}$, denoted by $\{E_n\}_{conv.}$, is the space $E = \bigcup_{n=1}^{\infty} E_n$, equipped with the finest locally convex topology τ_ω having the property that for all i, $\tau_\omega|E_n$ is coarser than τ_n.

Such a convex hull E is called an $\mathcal{L}$*-space*; if every E_n is an $\mathcal{F}$-space, then E is called an $\mathcal{LF}$*-space*.

The above definition of $\mathcal{L}\mathcal{F}$-space agrees with the one in [10, p.66]. Many of the early theorems in [10] are valid for $\mathcal{L}$-spaces as well as for $\mathcal{L}\mathcal{F}$-spaces; this is true, for instance, of Proposition 2, the lemma and corollary following Proposition 2, Proposition 4, Corollary 1, Corollary 2, and Proposition 5. Furthermore, [20] shows that every $\mathcal{L}$-space is complete.

<u>Definition 15.2.</u> A sequence $\{A_n\}$ of locally m-convex algebras is called m-<u>compatible</u> if it is compatible and if every A_n is a two sided ideal in A_{n+1}. If the convex hull of an m-compatible sequence $\{A_n\}_{n=1}^{\infty}$ is locally m-convex, then we call it an $\mathcal{L}$-<u>algebra</u>; if furthermore every A_n is an $\mathcal{F}$-algebra, then we call the convex hull an $\mathcal{L}\mathcal{F}$-<u>algebra</u>.

Remark. If $\{A_n\}$ is an m-compatible sequence of locally m-convex algebras, we may always consider the finest locally m-convex topology τ_{ω}^{m} on $\bigcup_n A_n$ having the property that, for all n, $\tau_{\omega}^{m}|A_n$ is coarser than τ_n. Unfortunately, we are able to prove almost nothing about τ_{ω}^{m} unless it coincides with τ_{ω}, in which case we are in the situation of Definition 15.2.

In order that the concepts of $\mathcal{L}$-algebra and $\mathcal{L}\mathcal{F}$-algebra be useful, we need some conditions which tell us when the convex hull of an m-compatible sequence of locally m-convex algebras <u>is</u> an $\mathcal{L}$-algebra. Before obtaining such conditions in Theorem 15.4 below, we need

<u>Definition 15.3.</u> Let A be an algebra, and S a subset of A. Then $L_A(S)$, the <u>left annihilator</u> of S in A, is $\{x \in A | xS = 0\}$. Similarly for the right annihilator $R_A(S)$.

Remark. It is easily seen that $L_A(S)$ is a left ideal in A; if S is a left ideal in A, then $L_A(S)$ is a two-sided ideal in A. Similarly for $R_A(S)$.

Theorem 15.4. <u>Let $\{A_n\}$ be an m-compatible sequence of locally m-convex algebras. Then a sufficient condition for $\{A_n\}_{conv.}$ to be an $\mathcal{L}$-algebra is that, for every A_n, one of the following two conditions be satisfied.</u> (These conditions will be discussed after the proof of the theorem).

1) <u>The one-sided ideals $R_A(A_n)$ and $L_A(A_n)$ are regular in A.</u>

2) (α) <u>A_n satisfies condition (g) of Proposition 4.2.</u>

(β) <u>There exists a $p \geq n$ such that</u>

$$R_A(A_n) + A_p = L_A(A_n) + A_p = A.$$

Proof. It is clear that, if $\{A_{n_k}\}$ is a proper subsequence of A_n, then the convex hull of $\{A_{n_k}\}$ is the same as that of $\{A_n\}$. We may therefore assume, without loss of generality, that in (1) the units modulo $R_A(A_n)$ and $L_A(A_n)$ are in A_{n+1}, and that in (2) $p = n + 1$.

To prove the theorem, we must show that, given any convex and symmetric subset U of A whose intersection with any A_n is a τ_n-neighborhood of 0 in A_n, there exists an m-convex and symmetric subset V of U with the same property. So let U be given. In constructing V, all sets will be symmetric, and we shall not mention this fact explicitly any more.

Pick a closed m-convex neighborhood V_1' of 0 in A_2 such that $V_1' \subset U$. Let $V_1 = V_1' \cap A_1$. Then

a) $V_1' V_1 \subset V_1$, $V_1 V_1' \subset V_1$.

b) $V_1 \subset U$.

Suppose now that, for some n, we have an m-convex neighborhood V_n' of 0 in A_{n+1} and closed m-convex neighborhoods V_k of 0 in A_k for all $k \leq n$, such that

a) $V_n' V_k \subset V_k$, $V_k V_n' \subset V_k$ for all $k \leq n$

b) $V_n \cap A_{n-1} = V_{n-1}$ (if $n \geq 2$)

c) $V_n \subset U$.

We will construct an m-convex neighborhood V_{n+1}' of 0 in A_{n+2}, and a closed m-convex neighborhood V_{n+1} of 0 in A_{n+1}, such that the above conditions hold with n replaced by n+1. This will complete the proof, for, having completed the induction, the set $V = \bigcup_{n=1}^{\infty} V_n$ satisfies all our requirements.

Let W_{n+1}^L be the closure, in A_{n+2}, of $V_n' + L_{A_{n+2}}(A_n)$, and similarly for W_{n+1}^R. Let us show that W_{n+1}^L and W_{n+1}^R are neighborhoods of 0 in A_{n+2}; we need, of course, only do this for W_{n+1}^L. The proof will be different for the two conditions of the theorem.

<u>Condition 1</u>. Let $e \in A_{n+1}$ be a right unit for $L_A(A_n)$ in A. Then e is also a right unit for $L_{A_{n+2}}(A_n) [= L_A(A_n) \cap A_{n+2}]$ in A_{n+2}. If $x \in A_{n+2}$, and $xe \in V_n'$, then $x = xe + (x - xe) \in V_n' + L_{A_{n+2}}(A_n)$. Hence $W_{n+1}^L \supset V_n' + L_{A_{n+2}}(A_n) \supset \{x \in A_{n+2} | xe \in V_n'\}$, and the latter set is a neighborhood of 0 in A_{n+2}.

Condition 2. Our assumption implies that $A_{n+1} + L_{A_{n+2}}(A_n) = A_{n+2}$. Let $x \varepsilon A_{n+2}$; then $x = \lambda(y + z)$, where λ is a scalar, $y \varepsilon V'_n$, and $z \varepsilon L_{A_{n+2}}(A_n)$. Hence every element of A_{n+2} is a scalar multiple of some element of W^L_{n+1}. Since W^L_{n+1} is m-convex and symmetric, and closed in A_{n+2}, it follows from Condition 2 that W^L_{n+1} is a neighborhood of 0 in A_{n+2}.

Our next step is to show that $W^L_{n+1}V_k \subset V_k$ and $V_k W^R_{n+1} \subset V_k$ for all $k \leq n$; we will do this only for W^L_{n+1}. Now $[V'_n + L_{A_{n+2}}(A_n)]V_k = V'_n V_k + 0 \subset V_k$, and hence $W^L_{n+1}V_k \subset V_k$ by continuity and the fact that V_k is closed in A_k.

Now pick an m-convex neighborhood V'_{n+1} of 0 in A_{n+2}, such that $V'_{n+1} \subset W^L_{n+1} \cap W^R_{n+1} \cap U \cap X_{n+1}$ where X_{n+1} is any neighborhood of 0 in A_{n+2} such that $X_{n+1} \cap A_n \subset V_n$. The previous paragraph shows that

$$(a_1) \qquad V'_{n+1}V_k \subset V_k \quad \text{and} \quad V_k V'_{n+1} \subset V_k \quad \text{for all } k \leq n.$$

Let us show that $(V'_{n+1} \cap A_{n+1}) \cup V_n$ is idempotent: Since $V'_{n+1} \cap A_{n+1}$ and V_n are both idempotent, and by (a_1) with $k = n$, we have $[(V'_{n+1} \cap A_{n+1}) \cup V_n]^2 \subset (V'_{n+1} \cap A_{n+1})^2 \cup V'_{n+1}V_n \cup V_n V'_{n+1} \cup V_n^2 \subset (V'_{n+1} \cap A_{n+1}) \cup V_n$.

Now let Y_{n+1} be the convex hull of $(V'_{n+1} \cap A_{n+1}) \cup V_n$, and let V_{n+1} be the closure of Y_{n+1} in A_{n+1}. The previous paragraph and Lemma 1.3 imply that Y_{n+1} is m-convex, and hence so is V_{n+1} by Lemma 1.4(b). Also $V'_{n+1}Y_{n+1} =$ convex hull of $V'_{n+1}[(V'_{n+1} \cap A_{n+1}) \cup V_n] \subset$ convex hull of $(V'_{n+1} \cap A_{n+1}) \cup V_n \subset V_{n+1}$, and similarly $Y_{n+1}V'_{n+1} \subset V_{n+1}$. Hence by continuity,

$$(a_2) \qquad V'_{n+1}V_{n+1} \subset V_{n+1}, \text{ and } V_{n+1}V'_{n+1} \subset V_{n+1}.$$

Since $V'_{n+1} \cap A_{n+1}$ is convex in A_{n+1}, and $(V'_{n+1} \cap A_{n+1}) \cap A_n \subset V_n$, the proof of the lemma following Proposition 2 of [10] shows that $Y_{n+1} \cap A_n = V_n$. Since V_n is closed in A_n, it follows that

(b) $$V_{n+1} \cap A_n = V_n .$$

Finally, since $V'_{n+1} \subset U$ and $V_n \subset U$, we have

(c) $$V_{n+1} \subset U .$$

By combining (a_1) and (a_2) to form (a), we have shown that V'_{n+1} and V_{n+1} satisfy all the required conditions.

Q.E.D.

Let us make some observations concerning the conditions of the theorem.

Condition 1. It is easy to see that $R_A(A_n)$ is regular if and only if either $R_A(A_n) = A$ or A_n has a right identity in A. (An element $e \varepsilon A$ is a right identity for A_n if $xe = x$ for all $x \varepsilon A_n$.)

The proof of the theorem shows that we may replace Condition 1 by the following weaker condition

(1') For some indices p and q, with $n \leq p < q$, there exist continuous projections r and ι from A_q into A_p such that, for any $x \varepsilon A_q$, $r(x) - x \varepsilon R_A(A_n)$ and $\iota(x) - x \varepsilon L_A(A_n)$.

Condition 1 and Condition 2: Condition 1 implies condition 2(β).

Both conditions are satisfied if all ring products are 0; in this case it is, of course, a priori obvious that the conclusion of the theorem is true.

Condition 2 (α): This is certainly satisfied if A_n is an $\mathcal{F}$-algebra.

Let us now give some examples of $\mathcal{L}$-algebras and $\mathcal{LF}$-algebras.

Example 15.5. Let $\{B_k\}$ be a sequence of $\mathcal{F}$-algebras, and let $A_n = B_1 \oplus \dots \oplus B_n$, with the product topology. Here condition 1' is satisfied. It is, moreover, easy to see directly that $\{A_n\}_{conv.}$ is an $\mathcal{LF}$-algebra.

Example 15.6. A_n is the Banach algebra of continuous complex-valued functions on the real line which vanish outside the closed interval [-n, n], with the "sup." norm. Here Conditions 1 and 2 are both satisfied. The resulting $\{A_n\}_{conv.}$ is the algebra of Example 3.5.

Example 15.7. Let A_n be the algebra of Example 3.4, with I = [-n, n].

Here Conditions 1 and 2 both apply. The convex hull of $\{A_n\}$ is the space $\mathfrak{D}$ of Schwartz [25, p.25 and p.67]. We therefore conclude that $\mathfrak{D}$ is not only an $\mathcal{LF}$-space, but also an $\mathcal{LF}$-algebra, (see Example 3.6).

Let us now look at the properties of $\mathcal{L}$-algebras. In the first place, they are complete, locally m-convex algebras, and the whole theory of such algebras which we have developed in this paper applies to them. Furthermore, they often inherit some of the special properties of the generating algebras. We have already seen one example of this in Lemma 12.3(d); another example is furnished by the following

Proposition 15.8. If $A = \{A_n\}_{conv.}$ is a commutative $\mathcal{L}$-algebra, and if each A_n is a Q-algebra, then A is a Q-algebra.

Proof. By Proposition 13.5, we need only show that $\mathcal{S}(A)$ is a neighborhood of 0. Since $\mathcal{S}(A)$ is m-convex (Lemma 13.2), we need only show that $\mathcal{S}(A) \cap A_n$ is a neighborhood of 0 in A_n for all n. But, since A_n is an ideal in A, $\mathcal{S}(A) \cap A_n = \mathcal{S}(A_n)$, and $\mathcal{S}(A_n)$ is a neighborhood of 0 in A_n by Proposition 13.5.

Q.E.D.

In particular, it follows that the space $\mathfrak{D}$ of [25] is a Q-algebra. This result could also have been obtained by using Theorem 13.6.

It would take us too far afield to systematically collect all the results of the above type. We will therefore conclude this section with the following

Proposition 15.9. Let A be an $\mathcal{L}$-algebra, generated by the sequence of locally m-convex algebras $\{A_n\}$. Then $\mathcal{M}^-(A) = \bigcup_n(\mathcal{M}^-(A) - (A_n)^o)$; for each n, there is a homeomorphism ϕ_n from $\mathcal{M}^-(A_n)$ onto $\mathcal{M}^-(A) - (A_n)^o$ such that $[\phi_n(f)](x) = f(x)$ for every $f \in \mathcal{M}^-(A_n)$ and $x \in A_n$. If, for every n, A_n is an $\mathcal{F}$-algebra and $\mathcal{M}(A)$ is second countable, then $\mathcal{M}^-(A)$ is second countable and locally compact.

Proof. That $\mathcal{M}^-(A) = \bigcup_n(\mathcal{M}^-(A) - (A_n)^o)$ is obvious. The existence of the homeomorphisms ϕ_n follows from Proposition C.3. If, for every n, A_n is an $\mathcal{F}$-algebra and $\mathcal{M}(A_n)$ is second countable, then every $\mathcal{M}(A_n)$ is second countable and locally compact (Lemma 6.2(c) and [1, p.486]) and so are the open subsets $\mathcal{M}^-(A_n)$. But then $\mathcal{M}^-(A)$ is the union of countably many open subsets each of which is second countable and locally compact, and therefore $\mathcal{M}^-(A)$ is also second countable and locally compact. Q.E.D.

Appendix A. Linear transformations between locally convex spaces.

Throughout this appendix, E and F will denote locally convex spaces; $\{U_i\}$ and $\{V_i\}$ will denote collections of convex and symmetric neighborhoods of the origin in E and F respectively, indexed by the same index set. For the notation used below, see Notation 2,6 and Convention 2.2.

Definition A.1.

a) $\mathcal{L}\{U_i, V_i\}$ will denote $\{T: E \rightarrow F \mid T \text{ linear}, T(U_i) \subset \lambda_{i,T} V_i$ for all i, where $\lambda_{i,T}$ is a scalar depending on i and T $\}$.

b) For each $T \in \mathcal{L}\{U_i, V_i\}$ and each i, the continuous linear transformation $T_i: E_i \rightarrow F_i$ is defined by "$T_i(x) = (T(x))_i$". (T_i is clearly well defined and continuous.)

In Theorem A.2 below, we will study an element T of $\mathcal{L}\{U_i, V_i\}$ by means of the corresponding T_i's. The theorem could be generalized to one about transformations between uniform spaces, but it does not seem worthwhile to do so here.

Theorem A.2. Suppose that $T \in \mathcal{L}\{U_i, V_i\}$. Then

a) If $\{V_i\}$ is a sub-base for F, then T is continuous.

b) If T_i is 1-1 for all i, and if $\{U_i\}$ is a sub-base for E, then T is 1-1.

c) If T_i is 1-1 and open for all i, and if $\{U_i\}$ is a sub-base for E, then T is 1-1 and open.

d) If $T_i(E_i)$ is dense in F_i for each i, and if $\{V_i\}$ is a base for F_i, then T(E) is dense in F.

e) If T_i is an isomorphism from E_i onto a dense subset of F_i for each i, if $\{U_i\}$ is a sub-base for E, if $\{V_i\}$ is a base for F, and if E is complete, then T is an isomorphism from E onto F.

Proof.

a) This is obvious.

b) We must show that the kernel of T is 0. Suppose that $x \neq 0$. Then $x \notin U_i$ for some i, and hence $x_i \notin \pi_i(U_i)$ for this i. Hence $x_i \neq 0$, so $(T(x))_i = T_i(x_i) \neq 0$, and therefore $T(x) \neq 0$.

c) For each i, T_i is open and therefore there exists an $\epsilon_i > 0$ such that $T_i(\pi_i(U_i)) \supset T_i(E_i) \cap \epsilon_i \pi_i(V_i)$. It is clearly sufficient to show that $T(U_i) \supset T(E) \cap \epsilon_i V_i$. Let $y \varepsilon T(E) \cap \epsilon_i V_i$; then $y = T(x)$ for some $x \varepsilon E$, so $y_i = T_i(x_i)$. But $y_i \varepsilon T_i(E_i) \cap \epsilon_i \pi_i(V_i)$, so $y_i = T_i z_i$ for some $z_i \varepsilon \pi_i(U_i)$. Since T_i is 1-1, it follows that $x_i = z_i \varepsilon \pi_i(U_i)$, whence $x \varepsilon U_i$.

d) Let $y \varepsilon F$. We need only show that for every i there exists an $x \varepsilon E$ such that $T(x) \varepsilon y + V_i$. Fix i. By assumption, there exists an $x \varepsilon E$ such that $(T(x))_i = T_i(x_i) \varepsilon y_i + V_i$. Hence $(T(x) - y)_i \varepsilon V_i$, whence $T(x) - y \varepsilon V_i$, so $T(x) \varepsilon y + V_i$.

e) By (a) and (c), T is an isomorphism into, and, since E is complete, T(E) is complete and hence closed in F. By (d), T(E) is dense in F, and therefore T(E) = F.

Q.E.D.

We turn to another matter, We will topologize $\mathcal{L}\{U_i, V_i\}$ in Definition A.3 below, and then proceed to derive some properties of this topology. If E = F and $\{U_i\} = \{V_i\}$ is a sub-base for E, then $\mathcal{L}\{U_i, V_i\}$, equipped with this topology, turns out to be a locally m-convex algebra (Proposition 2.4(j)). For a partial converse, see Proposition 4.4.

We shall assume throughout the remainder of this section that $\{V_i\}$ is a sub-base for F. It follows by Theorem A.2(a) that $\mathcal{L}\{U_i, V_i\} \subset \mathcal{L}(E, F)$. ($\mathcal{L}(E, F)$ is [10] the space of continuous linear transformations from E to F.)

Definition A.3. The topology $\tau\{U_i, V_i\}$ on $\mathcal{L}\{U_i, V_i\}$ is generated by taking as a sub-base for $\mathcal{L}\{U_i, V_i\}$ the collection of sets $\{T \varepsilon \mathcal{L}\{U_i, V_i\} \mid T(U_{i_0}) \subset V_{i_0}\}$, as i_0 ranges over the index set.

It is easy to verify that $\tau\{U_i, V_i\}$ is a locally convex topology on $\mathcal{L}\{U_i, V_i\}$ which is finer than the topology of uniform convergence on the bounded subsets of E. Let us study this topology.

Proposition A.4. *If F is complete, then so is* $\tau\{U_i, V_i\}$.

Proof. Let $\{T_\mu\}$ be a $\tau\{U_i, V_i\}$-Cauchy directed set in $\mathcal{L}\{U_i, V_i\}$. Then $\{T_\mu(x)\}$ is Cauchy in F for every $x \varepsilon B$, whence $T_\mu(x) \longrightarrow y \varepsilon F$. Define $T: E \longrightarrow F$ by "$T(x) = y$"; it is clear that T is linear, and that $T_\mu(x) \longrightarrow T(x)$ for every $x \varepsilon E$. We must show that

a) $T \varepsilon \mathcal{L}\{U_i, V_i\}$

b) $T_\mu \longrightarrow T$ in the topology $\tau\{U_i, V_i\}$.

Pick an index i, and any scalar λ_i. Since $\{T_\mu\}$ is $\tau\{U_i, V_i\}$-Cauchy, there exists a μ_o, such that $(T_\mu - T_\nu)U_i \subset \lambda_i V_i$ whenever $\mu, \nu \geq \mu_o$. Hence it follows that $(T_\mu - T)U_i \subset (\lambda_i + 1)V_i$ for $\mu \geq \mu_o$. Now $T_{\mu_o}(U_i) \subset \kappa V_i$ for some scalar κ, so $T(U_i) \subset (\lambda_i + 1 + \kappa)V_i$, which proves (a). Furthermore, (b) now follows from (a) and the underlined assertion.

Q.E.D.

Lemma A.5. If the index set is countable, then $\tau\{U_i, V_i\}$ is metrizable.

Proof. For then $\tau\{U_i, V_i\}$ has a countable sub-base.

Q.E.D.

Appendix B. Ideals in dense subalgebras.

Definition B.1. A topological algebra A is normal if every proper, closed, regular, right (resp. left) ideal I in A is contained in a closed, regular maximal right (resp. left) ideal in A.

Definition B.2. Let B be a topological algebra. A dense subalgebra A of B is snug in B if, whenever J is a closed, regular, maximal right (resp. left) ideal in B, then $J \cap A$ is a regular maximal right (resp. left) ideal in A.

The above concepts play an important role in Section 9 (see Theorem 9.3-Theorem 9.8), and the main purpose of this appendix is to derive adequate sufficient conditions for them. We begin by deriving conditions for normality, one of which involves the concept defined in Definition B.2. But first we need two more definitions, which extend the concepts of completely continuous operator and completely continuous algebra from the normed case to the locally convex case.

Definition B.3. A linear transformation T from a locally convex space E to a locally convex space F is called completely continuous if the image under T of some open set in E is totally bounded in F.

Definition B.4. A locally convex topological algebra is completely continuous if, for every $x \in A$, the transformations R_x and L_x from A to A are completely continuous.

Conditions for normality may now be summarized in

Proposition B.5. <u>If A is a topological algebra, then any one of the following conditions is sufficient for A to be normal.</u>

a) A <u>is a Q-algebra (see Appendix</u> E).

b) A <u>is completely continuous.</u>

c) A <u>is a snug subalgebra of some normal algebra</u> B.

This proposition, as well as Theorem B.7 and Proposition B.8, will be proved at the end of this appendix (after Lemma B.17). The principal application of condition (c) will be to the case where A is normed algebra and B its completion; in this case, B is a Q-algebra and hence automatically normal (by (a)).

Let us now turn to conditions under which A is a snug subalgebra of B. We need

<u>Definition B.6.</u> A topological algebra is <u>factor-finite</u> if A/J is finite dimensional for every closed, regular, maximal one sided ideal J in A.

Remark. It follows from [3, Theorem 1] or [17, p.811] that every commutative, locally convex algebra with continuous quasi-inversion is factor finite (in fact "factor-one"). This is a special case of Proposition B.8 (a) below.

The significance of factor-finiteness for our purposes is

Theorem B.7. <u>If</u> B <u>is a factor-finite topological algebra, then any dense subalgebra of</u> B <u>is snug in</u> B.

We now turn to sufficient conditions for an algebra to be factor-finite.

Proposition B.8. <u>Either of the following conditions is sufficient for a topological algebra</u> B <u>to be factor-finite.</u>

a) B <u>is a locally convex algebra with continuous quasi-inversion which satisfies a polynomial identity.</u>

b) B <u>is completely continuous.</u>

In applying the conditions of Proposition B.8, the following proposition, whose proof we leave to the reader, is sometimes useful.

Proposition B.9. <u>The property of satisfying a polynomial identity (being completely continuous) is maintained when passing to a subalgebra, the completion, or a finite cartesian product.</u>

We now prove some lemmas in preparation for the proofs of Proposition B.5, Theorem B.7, and Proposition B.9. Most of these lemmas have some independent interest, and some of them are stated with more generality than is necessary for our purposes.

Throughout Lemmas B.10, B.11 and B.13, $\bar{A}$ will denote a topological algebra, and A a dense subalgebra of $\bar{A}$.

The following lemmas are stated for right ideals, but it is clear that they remain true for left ideals (or even two-sided ideals in Lemmas B.10, B.11 and B.13).

Lemma B.10. *Let $\bar{A}$ be a topological algebra with dense subalgebra A. If K is a right ideal in A, and if $\bar{K}$ is the closure of K in $\bar{A}$, then $\bar{K}$ is a right ideal in $\bar{A}$. If K is regular in $\bar{A}$, then $\bar{K}$ is regular in $\bar{A}$; in fact, if e is a left unit in A modulo K, then e is also a left unit in $\bar{A}$ modulo $\bar{K}$.*

Proof. Since $xA \subset K$ for all $x \in K$, we have $x\bar{A} \subset \bar{K}$ for all $x \in K$, by the continuity of left multiplication. Hence $Kx \subset \bar{K}$ for all $x \in \bar{A}$, and therefore $\bar{K}x \subset \bar{K}$ for all $x \in \bar{A}$, by the continuity of right multiplication. But this says that $\bar{K}$ is a right ideal in $\bar{A}$.

If K is regular in A, and if e is a left unit in A modulo K, then $(e - 1)A \subset K$. Hence $(e - 1)\bar{A} \subset \bar{K}$, by the continuity of left multiplication, whence e is a left unit in $\bar{A}$ modulo $\bar{K}$.

Q.E.D.

Lemma B.11. *Let $\bar{A}$ be a topological algebra with dense subalgebra A. If K is a proper, closed, regular, right ideal in A, and if e is a left unit modulo K in A, (and if $\bar{A}$ is normal), then there exists a (closed) regular, maximal, right ideal J in $\bar{A}$ such that $K \subset J$ and such that e is a left unit modulo J in $\bar{A}$.*

Proof. Let $\bar{K}$ be the closure of K in $\bar{A}$. By Lemma B.10, $\bar{K}$ is a regular, right ideal in $\bar{A}$, and e is a left unit modulo $\bar{K}$ in $\bar{A}$. Since K is closed and proper in A, $\bar{K}$ is closed and proper in $\bar{A}$. Hence there exists a (closed) regular, maximal right ideal J in $\bar{A}$, such that $\bar{K} \subset J$, and such that e is a left unit modulo J in $\bar{A}$. It is clear that J has all the required properties.

Q.E.D.

Lemma B.12. Let B be an algebra, C a subalgebra of B, and K a regular right ideal in B such that $K + C = B$. Then $K \cap C$ is a regular right ideal in C, and there exists an $e \in C$ which is a left unit modulo K in B and modulo $K \cap C$ in C.

Proof. Let $f \in B$ be a left unit modulo K in B; then $f = e + k$, with $e \in C$ and $k \in K$. Now $(e - 1)B = (f - 1)B - kB \subset K$, so e is a left unit modulo K in B. Also $(e - 1)C \subset K \cap C$, so e is a left unit modulo $K \cap C$ in C, and hence $K \cap C$ is regular in C.

Q.E.D.

The above lemmas hold, with identical proofs, for arbitrary topological rings with operators. The next lemma, however, is meaningful only for algebras. Throughout its proof, we will tacitly use the fact that a finite dimensional topological linear space over the complex numbers (with multiplication by scalars jointly continuous) is isomorphic to a complex Euclidean space.

Lemma B.13. Let $\overline{A}$ be a topological algebra with dense subalgebra A. Let J be a proper closed right ideal in $\overline{A}$ such that the factor space $\overline{A}/J$ is finite dimensional (i.e., J has finite deficiency in $\overline{A}$). Then $J \cap A$ is a proper closed right ideal in A, and

(1) $A + J = \overline{A}$,

(2) $\overline{A \cap J} = J$.

If J is (maximal, regular, primitive) in $\overline{A}$, then $J \cap A$ is (maximal, regular, primitive) in A.

Proof. $J \cap A$ is clearly a closed right ideal in A. Since A is dense in $\overline{A}$ while J is not, we have $A \not\subset J$, and therefore $J \cap A$ is proper in A.

Consider now the canonical homomorphisms

$$\alpha : \overline{A} \longrightarrow \overline{A}/J$$

$$\beta : \overline{A} \longrightarrow \overline{A}/\overline{J \cap A}$$

$$\gamma : A \longrightarrow A/(J \cap A).$$

Observe first that $\overline{J \cap A} \cap A = J \cap A$, since $J \cap A$ is closed in A. We therefore have

(*) $\alpha(A)$, $A/J \cap A$, and $\beta(A)$ are algebraically isomorphic as vector spaces.

Since A is dense in $\bar{A}$, $\alpha(A)$ is dense in $\bar{A}/J$ and $\beta(A)$ is dense in $\bar{A}/\overline{J \cap A}$. But $\bar{A}/J$ is finite dimensional, so

$$(**) \qquad \alpha(A) = \bar{A}/J.$$

Hence $\alpha(A) = \alpha(\bar{A})$, from which (1) follows.

From (*) and (**) we see that $\beta(A)$ is finite dimensional, and hence

$$(***) \qquad \beta(A) = \bar{A}/\overline{J \cap A}.$$

Combining (*), (**), and (***), and remembering that $\overline{J \cap A} \subset J$, we get

(****) $\bar{A}/J$ and $\bar{A}/\overline{J \cap A}$ are algebraically isomorphic as vector spaces.

Since $\bar{A}/J$ is finite dimensional, this implies (2).

Now let J be maximal in $\bar{A}$, and suppose there were a right ideal I in A such that

$$(\dagger) \qquad J \cap A \subset I \subset A, \quad J \cap A \neq I \neq A.$$

To prove $J \cap A$ is maximal in A, we will show that I can be neither dense nor non-dense in A. If I is dense in A, then $\gamma(I)$ is dense in, but not equal to, the finite dimensional linear space $A/(A \cap J)$, which is impossible. If I is not dense in A, then K, the closure of I in A, satisfies

$$(\dagger\dagger) \qquad J \cap A \subset K \subset A, \quad J \cap A \neq K \neq A.$$

Hence $J = \overline{J \cap A} \subset \bar{K} \subset \bar{A}$, $J \neq \bar{K} \neq \bar{A}$, and since $\bar{K}$ is a right ideal in $\bar{A}$ (Lemma B.10), this contradicts the maximality of J.

If J is regular, then $(J \cap A)$ is regular by (1) and Lemma B.12.

If J is primitive, then $J \cap A$ is also primitive, since $A/J \cap A = A + J/J = \bar{A}/J$ by (1). Q.E.D.

Remark. The assertions about $J \cap A$ being regular and primitive follow from (1). The assertion about $J \cap A$ being maximal follows from (1) and (2).

Concerning finite dimensionality, let us observe that

Lemma B.14. <u>If B is an algebra, J a regular maximal right ideal in B, and P the right primitive ideal associated by J, then B/J is finite dimensional if and only if B/P is.</u>

Proof. This follows from the fact that B/P has a faithful representation as an indecomposable algebra of linear transformations on B/J.

Q.E.D.

Lemma B.15. <u>Let E be a topological linear space, and H a closed subspace of E such that E/H is finite dimensional. Then any subspace of F of E which contains H is closed.</u>

Proof. Let α be the canonical projection of E onto E/H. Now $\alpha(H)$ is certainly closed in the finite dimensional space E/H, and hence $H = \alpha^{-1}(\alpha H)$ is closed in E.

Q.E.D.

Lemma B.16. <u>If J is a closed, regular, right ideal in a completely continuous algebra B, then B/J is finite dimensional.</u>

Proof. Let J be a closed, regular, right ideal in B, and let e be a left unit modulo J in B. Then e, acting as the identity transformation on the topological linear space B/J, is completely continuous. Hence B/J has a totally bounded neighborhood of the origin, and is therefore finite dimensional.

Q.E.D.

Lemma B.17. <u>If B is a locally convex algebra with continuous quasi-inversion which satisfies a polynomial identity, and if P is a closed primitive ideal in B, then B/P is finite dimensional.</u>

Proof. Clearly B/P is a primitive algebra satisfying the same identity as B. Hence, by [16, Theorem 1], B/P is finite dimensional over its center. But this center is a locally convex division algebra with continuous inversion, and is therefore isomorphic to the algebra of complex numbers [3, Theorem 1].

Q.E.D.

We are now ready to prove our main results.

<u>Proof of Proposition</u> B.5.

a) This follows immediately from Zorn's lemma and Lemma E.4.

b) This follows from Lemma B.16 and Lemma B.15.

c) Let I be a closed, regular, right ideal in A. By Lemma B.11, there exists a closed, regular, maximal right ideal J in $\bar{A}$ which contains I. Now $J \cap A$ clearly contains I, and $J \cap A$ is closed regular maximal right ideal in A by assumption.

Q.E.D.

Proof of Theorem B. 7. This follows from Lemma B.13.

Proof of Proposition B.8.

a) This follows from Lemma B.17 and Lemma B.14.

b) This follows from Lemma B.16.

Q.E.D.

Appendix C. Extension of homomorphisms. Throughout this section, A will denote an algebra, and I a two-sided ideal in A.

In our first theorem, we slightly improve on [6, Theorem 1] by removing an unnecessary assumption.

Theorem C.1. *Let B be an algebra with unit. Then any homomorphism f from I onto B can be uniquely extended to a homomorphism $\hat{f}$ from A to B. If A is a topological algebra, and if f is continuous, then $\hat{f}$ is continuous.*

Proof. Pick $e \in I$ such that $f(e)$ is the identity element of B. If $x \in A$, then $f(xe) = f(exe)$. Now define $\hat{f}: A \longrightarrow B$ by

$$\hat{f}(x) = f(xe) = f(ex).$$

It is clear that $\hat{f}$ is a linear transformation which extends f. Furthermore, for any x and y in A, $\hat{f}(xy) = f(xye) = f(exye) = f(ex)f(ye) = \hat{f}(x)\hat{f}(y)$, so $\hat{f}$ is in fact a homomorphism.

To see that $\hat{f}$ is unique, let h be any homomorphism from A to B which extends f. Then, for any $y \in A$, $h(y) = h(y)h(e) = h(ye) = f(ye) = \hat{f}(y)$.

The last assertion of the theorem is obvious from the definition of $\hat{f}$. Q.E.D.

Just as in [6], Theorem F.1 yields the following interesting corollary, which will not, however, be applied in this paper.

Corollary C.2. *If J is a regular, two-sided ideal in I, then there exists a unique two-sided ideal J' in A having the following properties:*

a) $J' \cap I = J$

b) $J' + I = A$.

The ideal J' is regular; in fact, if e is a unit modulo J in I, then e is a unit modulo J' in A. If J is maximal in I, then J' is maximal in A, and moreover J' *is the only maximal two-sided ideal in* A *satisfying* (a). *Finally, if* A *is a topological algebra and if* J *is closed in* I, *then* J' *is closed in* A.

In the following proposition, which will be used in the proof of Proposition 15.9, we will apply Theorem C.1 to the case where A is a topological algebra, and B is the algebra of complex numbers.

Proposition C.3. *Let* A *be a topological algebra, and* I *a two-sided ideal in* A. *Define* $\phi: \mathcal{M}^-(I) \longrightarrow \mathcal{M}^-(A)$ *by* "$\phi(f) = \hat{f}$" (*see Theorem* C.1). *Then* ϕ *is a homeomorphism from* $\mathcal{M}^-(I)$ *onto an open subset of* $\mathcal{M}^-(A)$.

Proof. It is clear that ϕ is 1-1 and open. Let us show that ϕ is continuous. For $x \in A$, $\approx{x}$ will denote the complex valued function on $\mathcal{M}^-(A)$ generated by x, and if $x \in I$, $\tilde{x}$ will denote the function on $\mathcal{M}^-(I)$ generated by x. Let $f_o \in \mathcal{M}^-(I)$, and pick $e_o \in I$ such that $f_o(e_o) = 1$. Since e_o is continuous on $\mathcal{M}^-(I)$, we can find a neighborhood U_o of f_o in $\mathcal{M}^-(I)$ such that $e_o(f) \neq 0$ for $f \in U_o$. Now to show that ϕ is continuous at f_o, we must show that, for every $y \in A$, the mapping $f \longrightarrow \overset{\approx}{y}(\hat{f})$ is continuous on $\mathcal{M}^-(I)$, and it is sufficient to show that it is continuous on U_o. Now, for $f \in U$,

$$\overset{\approx}{y}(\hat{f}) = \frac{\overset{\approx}{y}(\hat{f})\tilde{e}_o(f)}{\tilde{e}_o(f)} = \frac{\hat{f}(y)\hat{f}(e_o)}{\tilde{e}_o(f)} = \frac{\hat{f}(ye_o)}{\tilde{e}_o(f)} = \frac{f(ye_o)}{\tilde{e}_o(f)} = \frac{\widetilde{ye}_o(f)}{\tilde{e}_o(f)},$$

which is certainly a continuous function of f on U_o.

We have shown that ϕ is a homeomorphism into. To see that $\phi(\mathcal{M}^-(I))$ is open in $\mathcal{M}^-(A)$, we need only observe that $\mathcal{M}^-(A) - \phi(\mathcal{M}^-(I)) = \bigcap_{x\in I} \{g \in \mathcal{M}^-(A) \mid g(x) = 0\}$, and this last set is certainly closed.

Q.E.D.

Appendix D. Generating collections of subsets and completeness of function algebras. The contents of this section are closely related to Section 2 of [5].

Definition D.1. Let X be a topological space, and $\mathcal{S}$ a collection of subsets of X whose union is X. Then $\mathcal{S}$ *generates* X if a subset of X is open in X provided its intersection with every S ε $\mathcal{S}$ is open in S in the relative topology on S.

Remark. If $\mathcal{S}$ generates X, and if $\mathcal{T} \supset \mathcal{S}$, then $\mathcal{T}$ generates X.

Remark. If the collection of compact subsets of X generates X, then Hurewicz calls X a k-*space* (see, for instance [11]).

Let us now give two examples of generating collections.

Example D.2. If X is first countable, then the collection of all convergent sequences in X, with limit attached, generates X. Hence X is a k-space.

Example D.3. If X is first countable or locally compact, then the collection of compact subsets of X generates X (i.e., X is a k-space) [11].

Lemma D.4. *If X is a topological space with generating collection $\mathcal{S}$ of subsets, and if f is a function from X to some topological space Y whose restriction to every S ε $\mathcal{S}$ is continuous on S, then f is continuous on X.*

Proof. Let U be open in Y. Then $f^{-1}(Y) \cap S$ is open in S for each S ε $\mathcal{S}$, and hence Y is open in X by Definition D.1.

Q.E.D.

The most important application, for our purposes, of the concept of a generating collection, is

Lemma D.5. *If X is a topological space with generating collection of subsets $\mathcal{S}$, if Y is a uniform space (say a topological group), and if C(X, Y) is the collection of continuous functions from X to Y, then C(X, Y) is complete in the uniform structure of uniform convergence on the elements of $\mathcal{S}$.*

Proof. It follows from [5, Theorem 1 and Proposition 2] that $\mathcal{F}$(X, Y), the space of *all* functions from X to Y, is complete in this uniform structure, so we need only show that C(X, Y) is closed in $\mathcal{F}$(X, Y). Now if f ε $\mathcal{F}$(X, Y) is in the closure of C(X, Y), then clearly, for every

$S \in \mathcal{S}$, the restriction $f|S$ of f to S is the uniform limit of functions which are continuous on S, and therefore $f|S$ is continuous on S. Hence $f \in C(X, Y)$ by Lemma D.4.

Q.E.D.

Appendix E. Q-algebras. Following [15, p.155], we make the following definition.

Definition E.1. A topological algebra A is a *Q-algebra* if the set of q.r. elements of A is open in A.

Lemma E.2. *A topological algebra A is a Q-algebra if and only if the set S of q.r. elements of A has an interior.*

Proof. One half of the Lemma is trivial. To prove the other half, we proceed as follows: For any $x \in A$, define the continuous functions $\phi_x : A \longrightarrow A$ and $\psi_x : A \longrightarrow A$ by "$\phi_x(y) = x \circ y$", "$\psi_x(y) = y \circ x$". Now if $x \in A$, then any element of $\phi_x^{-1}(S)$ (resp. $\psi_x^{-1}(S)$) is r.q.r. (l.q.r.); hence, if x and y are in A, then $\phi_x^{-1}(S) \cap \psi_x^{-1}(S) \subset S$. Suppose now that x is an interior point of S, and let z be any point of S. Then $\phi_{x \circ z'}(z) = \psi_{z' \circ x}(z) = x$, and therefore z is an interior point of $\phi^{-1}_{x \circ z'}(S) \cap \psi^{-1}_{z' \circ x}(S) \subset S$.

Q.E.D.

The basic properties of Q-algebras are summarized in the following two lemmas. The first of these follows immediately from the definition, while the second is part of [26, Theorem 1.6].

Lemma E.3. *Every element of a Q-algebra has a compact spectrum.*

Lemma E.4. *Every regular, maximal ideal (right, left, or two-sided) in a Q-algebra is closed.*

It is well known that every Banach algebra is a Q-algebra. There are, however, non-normed locally m-convex algebras which are Q-algebras; Example 3.4, Example 3.5, and Example 3.6 provide examples of such algebras. For theorems about Q-algebras in the main part of this paper, see Proposition 5.10 and Section 13.

Bibliography

1. Arens, R.F. A topology for spaces of transformations, Ann. of Math., (2) 47, 480-495 (1946).

2. Arens, R.F. The space L^ω and convex topological rings, Bull. Amer. Math. Soc. 52, 931-935 (1946).

3. Arens, R.F. Linear topological division algebras, Bull. Amer. Math. Soc. 53, 623-630 (1947).

4. Banach, S. Théorie des operations linéaires, Monografje Matematyczne, Warsaw, 1932.

5. Bourbaki, N. Topologie Générale Ch.X, Actualités Scientifiques et Industrielles, No. 1084, Paris, 1949.

6. Buck, R.C. Extensions of homomorphisms and regular ideals, J. Ind. Math. Soc. 14, 156-159 (1950).

7. Cartan, H. Ideaux et modules de fonctions analytiques de variables complexes, Bull. Soc. Math. France 78, 29-64 (1950).

8. Cartan, H. and Thullen, P. Zur Theorie der Singularitäten der Funktionen mehrerer komplexen Veränderlichen. Regularitäts und Konvergenzbereiche, Math. Ann. 106, 617-647(1932).

9. Dieudonné, J. La dualité dans les espaces vectoriels topologiques, Ann. Ecole Norm. 59, 107-140 (1942).

10. Dieudonné and Schwartz, L. La dualité dans les espaces $(\mathcal{F})$ et $(\mathcal{L}\mathcal{F})$, Ann. Inst. Fourier 1, 61-101 (1950).

11. Gale, D. Compact sets of functions and function rings, Proc. Amer. Math. Soc. 1, 303-308 (1950).

12. Gelfand, I. Normierte Ringe, Rec. Math. (Math. Sbornik) N.S.9, 3-24 (1941).

13. Graves, L. Theory of functions of real variables, McGraw-Hill, New York and London, 1946.

14. Hewitt, E. Rings of real valued continuous functions I, Trans. Amer. Math. Soc. 64, 45-99 (1948).

15. Kaplansky, I. Topological Rings, Amer. J. Math. 69, 153-183 (1947).

16. Kaplansky, I. Rings with a polynomial identity, Bull. Amer. Math. Soc. 54, 575-580 (1948).

17. Kaplansky, I. Topological Rings, Bull. Amer. Math. Soc. 45, 809-826 (1948).

18. Kaplansky, I. Normed Algebras, Duke Math. J. 16, 399-418(1949).

19. Kaplansky, I. Locally compact rings II, Amer. J. Math. 73, 20-24 (1951).

20. Köthe, G. Uber die Vollständigkeit einer Klasse lokalkonvexer Räume, Math. Zeit. 52, 627-630 (1949-1950).

21. Mackey, G.W. On finite-dimensional linear spaces, Trans. Amer. Math. Soc. 57, 155-207 (1945).

22. Mackey, G.W. On convex topological linear spaces, Trans. Amer. Math. Soc. 60, 520-537 (1946).

23. Newburg, J.D. The variation of spectra, Duke Math. J. 18, 165-177 (1951).

24. Rickart, C.E. The uniqueness of norm problem in Banach algebras, Ann. of Math. 51, 615-628 (1950).

25. Schwartz, L. Théorie des Distributions, Tome I, Actualités Scientifiques et Industrielles, No. 1091, Paris, 1950.

26. Segal, I.E. The group algebra of a locally compact group, Trans. Amer. Math. Soc. 61, 69-105 (1947).

27. Stone, M.H. The generalized Weierstrass approximation theorem, Mathematics Magazine 22, 167-183 and 237-254(1947-1948).

28. Weil, A. L'integration dans les groupes topologiques et ses applications, Actualités Scientifiques et Industrielles, No.869, Paris, 1940.

29. Helmer, O. Divisibility properties of integral functions, Duke Math. J . 6, 345-356 (1940).

30. Arens, R. A generalization of normed rings, to appear in Pacific J. Math. 2 (1952).